Helping Your Health with ENZYMES

CARLSON WADE

ARC BOOKS, INC.
New York

TO MY MOTHER

who discovered Nature's secrets
of health, enabling me
to enjoy a life of perpetual youth.

Published by ARC BOOKS, Inc.,
a division of Arco Publishing Company, Inc.
219 Park Avenue South, New York, N. Y. 10003
By arrangement with Parker Publishing Company, Inc.
Second Printing, January 1971
Copyright © 1966, by Parker Publishing Company, Inc.
All Rights Reserved
Standard Book Number 668-02131-4
Printed in U.S.A.

A DOCTOR'S FOREWORD

Optimal bodily health means that every cell in the body is performing its functions properly and adequately. These functions are metabolism, repair, and work.

Basically, good health depends upon the integrity of metabolic function. The metabolic function of each cell is determined and controlled by the presence of many different enzymes—all of which are proteins. The control of intracellular metabolism is, therefore, the control of the regulation of the types, amount and activities of these enzymes.

Normally, in the human body, the amount of some enzymes varies with certain nutritional and hormonal factors. Consequently, it is important that one have at all times an adequate protein diet accompanied by good digestion and assimilation.

Since your health is your most valuable asset, you cannot afford not to spend whatever time and energy is necessary to get at the truth underlying health and apply it to your own health problem. Collect all the evidence for and against every statement or opinion about diet; weigh each and come to your own conclusions. Check this conclusion with the opinion of men whom you feel you can trust.

Carlson Wade has written a good book regarding enzymes for this purpose. Use it as a basis for your own investigation and use of the impartial evidence he so clearly states in this book.

JONATHAN FORMAN, M.D.

WHAT THIS BOOK CAN DO FOR YOU

Wouldn't it be a wonderful feeling to awaken one bright morning and discover that you have no colds, no headaches, that there is no ache in your back, that you feel calm and relaxed? What an experience to be filled with energy, good cheer, hopeful anticipation to greet the day ahead of you! Yes, good mental and physical health is a precious treasure.

But how often do you wake up, following a night of sleepless tossing and turning, feeling tired, achey, foggy, with little enough desire to do anything? The new day ahead is just another brief stretch of time in which you will push yourself to do routine work. The blessings of good health are denied to you.

Must this be so? Decidedly not! You *can* enjoy happy health! You *can* wipe away the effects of the years of being only half-alive! You *can* taste the sweet nectar of Nature's own fountain of youth! It is all up to you—and the foods that you eat!

You are not alone in this quest for better health and longer life. You have, at your disposal, close to 700 different helpers. And these 700 individuals are not singular. No, indeed not. There are millions upon millions of each of these 700 helpers—and they are found in two places: right within your digestive system and in the foods that you eat.

These helpers are ENZYMES—the miracle health builders that give you life and health. Until recently, enzymes remained in the background. It was said that they serve the sole purpose of digesting foods and extracting the vitamins, minerals, proteins and other essential nutrients and then helping them become built into your body structure. True, this is a vital process. After all, without enzymes, life would perish.

But what about enzyme deficiencies? Just like with vitamin, mineral or protein deficiencies, the state of your mental and physical health is in peril. But scientists are now making new discoveries about these astonishing enzymes. For example, they have been used to dissolve scar tissue, to help eliminate the pressures and pains of a slipped disk, to reduce swelling conditions in certain body ailments. New discoveries have also found that since enzymes are needed for every single step in the metabolic process, a minute deficiency inevitably paves the way to illness and severe incapacitations.

A survey conducted by an outstanding authority tells us that for many years medical practice has been greatly concerned with the acute *infectious* diseases, but today a great problem is the *metabolic* diseases— those which are directly affected by what we eat! This authority emphasized that not only does half the population have some form of chronic disease, but only 13 per cent of the remainder are free from some type of physical defect. The most astonishing part of this report is the revelation that there are approximately 88,959,534 cases of chronic illness in our nation. It is estimated that 1 out of every 5 persons has one or more ailments—and the other 4 persons may have varying degress of ill health that are not serious enough to warrant medical attention—yet neither can they boast of being in topnotch physical or mental health.

The foods you eat and the nutrients extracted by your digestive system may well be the most valuable factors involved in building your health, and immunity to ailments.

Enzymes, as this book will show, have been given scant attention. *Yet, they are responsible for nearly every facet of life and health and far outweigh the importance of virtually any other nutrient.* Enzymes are part of all living things, from the cells in a weed, to the cells in your brain. The powerful influence of these minute compounds become significant when we learn that each body cell (you have millions of them) has at least 100,000 enzyme particles. Can a laboratory take a piece of celery or chicken drumstick and turn it into human parts? Of course not. Enzymes do this by means of digestion, extracting essentials which are then built into your body. Enzymes help give you warmth, energy, mentality, strength to blink your eye, run for a bus, figure a mathematical problem, or concoct a recipe—*enzymes have a part to keep your heart beating.*

Furthermore, enzymes rally to strike at diseases that have invaded the body. It has been observed that weak enzyme behavior may be responsible for leukemia, cancerous infections, mental disturbances, overweight, underweight, blood ailments, etc. In fact, you would find it difficult (if not impossible) to name a single internal or external body function that is possible without the prodding action of enzymes.

This book is particularly meant for the average man or woman who is aware of problems of deteriorating health and wants to help improve this situation.

This book is *different* from just about any other health book because it has the unique feature of enzymes as a means of building and maintaining health. Few books, if any, have touched on this subject to an appreciable level.

Other books have discussed vitamins, minerals, wheat germ, yogurt, lecithin, desiccated liver, safflower seed oil, etc. Few have discovered *how enzymes rule over all other nutrients.* Few books have explained what enzymes are, the foods in which they are located, how they function in health, how to super-charge your system with enzymes. In short, enzymes have been overlooked, and they may well be the "secret key" to glowing health.

Has this ever happened to you? Dinner is about to be served at the home of your business associate. You are the invited guest. Your host is served a steaming plate of tenderloin steak, mashed potatoes with gravy, a plate of raw and cooked vegetables; the rest of his dinner consists of fluffy dinner rolls with butter, cup of soup, a dessert of strawberry shortcake, or apple pie a la mode.

You are served a platter of plain boiled meat, potatoes and some soggy vegetables. Your dessert is apt to be simple sponge cake. Your excuse for such a meagre fare? "Sorry, wish I could eat that steak and everything, but it just wouldn't agree with me."

Why are others able to eat heartily, with gusto, and look the picture of health while you are denied the joys of feast-like meals—with all the trimmings?

The answer lies in enzymes—those built-in digestive aids that enable one person to eat everything that is placed before him and *enjoy* every mouthful—while another person suffers from gastric stomach upset, indigestion, heartburn, nausea and a score of ailments caused by malfunctioning of the digestive system.

This book is dedicated to better health via enzymes. Within reason, you should be able to eat almost anything if it is prepared reasonably well and in those combinations food may best be eaten. The entire secret of the so-called "cast-iron stomach" centers around digestive enzymes. Once you know how these enzymes work, you can use them to serve your health needs—and eat heartily and joyously as well.

If you are denying yourself a wide assortment of tempting foods because they "do not agree" with you, not only are you missing out on a lot of delicious taste satisfactions, but you are shortchanging your health. The very foods that give you an upset stomach may contain nutrients and substances *you need* to build health.

Few people are aware of enzymes that constitute the digestive juices which create abundant health, build resistance to disease, give you a new life and invigorated youthful appearance. By knowing the secrets of "fasting," the "trick" about adequate chewing, special breathing steps, the value of certain raw foods, the amazing way to build stomach health by practicing self-hypnotism just ten minutes before sitting down at the table, you can put life back into your years. Enzymes may well be in the spotlight in the quest for the fountain of youth because they can supercharge your mind and your body like nothing else can. WAKE UP AND LIVE—WITH ENZYMES! This book can show you how!

CONTENTS

What are enzymes? The hitherto undiscovered internal gushing springs of health building waters. How they perform digestion. How enzymes build strong bones, wrinkle-free skin, healthy muscles, youthful bloodstream. Enzymes needed to battle diseases. Where enzymes exist. Where enzymes come from. What happens when you are enzyme deficient. Each cell of billions of body tissues needs enzymes to trigger the vital body processes.

Topnotch health, long life, eternal youth all depend on smooth functioning digestion which is, in turn, completely "ruled" by enzymes. How does your digestive

*Extreme temperatures of heat and cold will destroy
enzymes or weaken their powers to render them
useless. How you may think your foods are enzyme
rich yet cooking processes have depleted vital stores.
How cooking is destructive to enzymes—yet you CAN
cook enzyme-rich foods and preserve much of the supply.
How to preserve enzymes when cooking.*

*Digestion is possible only in the presence of enzymes
which are needed to break down into digestible form
such essential foods as fat, starch, and protein. The
pouring forth of enzymes in the digestive tract is possible
when food is properly chewed and made ready to greet
these digestive powers. Digestion begins in the mouth.
How chewing can keep you slim even when you eat
starch or carbohydrates. Enzymes nullify many harmful
effects of acid foods, sweets, etc. Discover Fletcherism—
the way to "put teeth in your stomach." Hydrochloric
acid needed to digest meats is available when you devote
proper chewing time.*

*Fasting is an ancient remedy for modern ailments.
How fasting repairs internal organs, gives the digestive*

1

ENZYMES—YOUR "BUILT-IN" FOUNTAIN OF YOUTH

Suppose, one day, you complain of a chest pain. There are accompanying symptoms to indicate a possible heart attack. Is it true that a portion of your heart muscle is dying because a clot is blocking a coronary vessel, shutting off the vital blood supply?

Another day, you examine yourself carefully in the mirror and unhappily note that firm, clear and resilient skin has turned to wrinkled, patchy, greyish folds that scream premature aging.

Still another day, you partake of a delicious steak and mashed potatoes, feeling "full" for a few hours, then "heavy" for most of the night as indigestion ruins what should have been a delicious meal.

Your doctor may prescribe a new regimen for possible heart strain; he may suggest creams and lotions for aging skin. He may also administer medicine to ease discomfort following a full meal and suggest further that you go easy on the heavier types of foods because "you're not as young as you used to be."

But your doctor may also take a simple blood test and examine it for tiny amounts of chemical substances that will tell him much about your chest pain, your sick skin, your upset stomach. He will refer to this test as "hunting for molecules" so that he may be able to make an early and accurate diagnosis of what ails you.

If he finds that you have a deficiency of these molecules, which may be responsible for your ailments, he will be on the right track in helping you overcome the barriers of time. He will have discovered that you, like millions of others, are unaware of a hitherto secret fountain of youth—

right within your own body. This fountain issues forth internal gushing springs of health-building waters that have the power to restore, build, maintain vibrant youth and health. These waters also have the powers to destroy your health. Nature demands that this built-in fountain be properly tended and cared for if you are to reap the maximum benefits.

What are enzymes?

The waters issued within your digestive system are called *enzymes*. When examined, these enzymes tell your doctor whether you are healthy, ailing or in need of stimulation and invigoration—possible ONLY by proper diet and a simple fasting plan to be outlined in later chapters of this book.

Donald G. Cooley, writing in *Today's Health*, declares, "A drug that not only helps to diagnose disease, but also reduces black eyes and other swellings, cleans dirty wound surfaces like a gentle chemical scalpel, acts against inflammations, promotes healings, assists in delicate cataract surgery, aids digestion, dissolves clots, liquefies thick secretions that hinder breathing, and stops penicillin reactions, is too fantastic to be credible. Yet there is a class of substances that do all these things and more. *They are called enzymes.*

"We live with, and by, enzymes every instant of our lives. Your body at this moment seethes and simmers with millions of enzymatic reactions of which you are comfortably unaware. Without them you could not see these words, or understand them, or turn this page, or rise in annoyance when the phone rings, or enjoy your latest meal. It short, you would be a mere collection of inanimate atoms."

How do enzymes work?

The word enzyme is derived from the Greek *enzymos* which means "fermented" or "leaven"—to cause a change. Enzymes are regarded as catalysts—they have the power of causing an internal reaction without themselves being transformed or destroyed in the course of the process. Strictly speaking, an enzyme is an internal juice formed by the living tissues and cells of your digestive tract. Depending upon the task at hand, the enzyme flow may be very heavy or very light.

As an example of a catalyst, let us look at the flame on top of your stove, cooking food in the pot. The flame causes the food to undergo certain changes, yet the flame does not become part of the food. You do not eat the fire. You eat the food that has been cooked by the fire. The flame is a catalytic agent, or a catalyzer. Without its influential action, the food could not be cooked.

How many enzymes are there?

You do not have just a few enzymes. Your body has more than 700 different *types* of enzymes, each one performing a separate function. Of these 700-plus enzymes, you have millions upon millions of each one—found right within your digestive system and also in the foods that you eat.

When a single drop of blood is examined under the microscope, it shows at least 100,000 enzyme particles. These enzymes are spread throughout your entire body, from head to toe, helping to keep you alive, healthy, mentally and physically alert and as youthful as possible. When you have a deficiency of one or more of these vital enzymes, you run the risk of ill health.

Without enzymes, seeds could not sprout; wine and beer could not ferment; leaves could not change their color in autumn; tobacco could not be cured. Without enzymes, your food could not be digested to release valuable vitamins, minerals, amino acids needed to keep you alive and healthy. In addition, we can see the action of enzymes when they cause the ripening of green tomatoes or brown-flecked bananas into luscious red tomatoes and golden-hued bananas. Without the presence and action of enzymes, these changes could not be made.

Each enzyme acts upon a specific food; one will not substitute for the other. This means that a shortage or deficiency, or even absence of one single enzyme, can spell all the difference between life and health and death!

James S. McLester, M.D., in *Nutrition and Diet in Health and Disease*, has this to say: "As crystalline organic compounds, these materials are lifeless; as substances which have the property of increasing in the presence of living cells, they assume a property characteristic of living things."

Enzymes need to be activated into action if they are to serve you properly; otherwise, they become crystalline organic compounds in suspended animation and are actually worthless to you.

Your salivary glands, your pancreas, your stomach walls, your intestines contain the most valuable digestive enzymes; even after food has been changed by these enzymes into a form that can be shipped to all body cells, you must have still more enzymes in other cells to continue the job of changing the digested food into a substance that the body must have.

Enzymes are named in accordance with the food substance they work upon. For example, an enzyme that causes a change in the presence of phosphorus is called *phosphatase*. An enzyme that breaks down sugar (sucrose) into usable form is known as *sucrase*, and so forth.

Let us take a look at your various body systems to see how enzymes work to keep you alive and healthy.

Enzymes in your digestive system.

Your stomach is a distensible sac lying basically in the upper left region of your abdomen. It has the shape of the letter J, with the hook extending across the abdomen's midline to lie close to your liver's undersurface. The lining of your stomach is a prime source of glands which issue forth the vital enzymatic streams or internal fountains. *Pepsin* is a vital digestive enzyme which pours over the proteins of ingested food, splitting these proteins into usable amino acids. Without the presence of pepsin, protein could not be used by your body to build a healthy skin, a strong skeletal structure, a rich blood supply, a sparkle in your eyes, resistance against ailments, recovery from illness, strong muscles. Pepsin is the enzyme which ranks first in the entire system of digestion of proteins by splitting them into two parts known as proteoses and peptones.

Rennin is another valuable digestive enzyme which causes the coagulation of milk, changing its protein, *casein*, into a form that is usable by your body. Rennin releases the valuable mineral elements of milk so that calcium, phosphorus, potassium, iron, may then be used by your body to stabilize the water balance, strengthen your nervous system so that you can think clearly, build a rich red bloodstream, have strong teeth and bones. Without rennin, the vital nutrients in milk and dairy foods could not otherwise be released to provide their health-building functions.

Lipase is the next most valuable enzyme which serves the purpose of splitting fat which is then utilized to nourish your skin cells, cushion your body against bruises and blows, ward off the entrance of infectious virus cells and allergic conditions. While the small intestine is primarily concerned with fat digestion, it is in your digestive system where the "preliminaries" are made. Those of you who complain of stomach pains or indigestion after a meal of fat-containing foods, should be aware of the power of lipase which helps to partially digest fatty foods to prepare them for later enzyme action in the small intestine.

Salivary digestion.

The membranes of your mouth and pharynx (back of your throat, extending into the lower part leading into your windpipe and food tube) issue forth valuable enzymes which are needed to prepare food for swallowing and later digestion. Saliva contains the salivary enzyme called *ptyalin* (salivary amylase) which sparks the first step in digestive action. Starch foods—including breads, cakes, spaghetti, macaroni—are attacked by ptyalin to become broken down into what is called *maltose*. If you bolt down your food without giving time to proper chewing, then ptyalin is denied its rightful function.

Food then reaches the digestive system in a partially prepared form and renders utilization more difficult. In chewing, it is necessary to crush solid food by the power of the tongue and the grinding action of the teeth; when this is done, ptyalin is able to extract sugar from starchy foods and thereby provide you with a concentrated form of energy, alertness, and vigor.

Unchewed foods that are high in starch content will reach your gastric system in an unprepared state and the acid content in the stomach enzymes may counteract whatever ptyalin started to do. This leads to stomach pains, a feeling of heaviness and uncontrollable sleepiness after a meal. So—chew your foods if you want your enzymes to keep you healthy!

Intestinal digestion.

Enzymes perform their greatest jobs in your small intestine. Enzymes coming from your liver, pancreas and intestinal mucosa all combine to extract vital nutrients from food to be absorbed by your bloodstream. Liver cells issue *bile*, a brown or greenish-brown fluid. Bile is needed to emulsify fats, reducing all globules so there will be less tendency for them to combine with each other. Further, bile prepares these extracted nutrients from fats to be absorbed into your bloodstream to be carried to all parts of your body.

Once again, we see how a deficiency of bile or other enzymes concerned with intestinal digestion, render fat a "difficult to absorb" food.

Pancreatic juices.

Your pancreas is a long and large organ found behind the lower part of the stomach. Its major task is to manufacture enzymes which are passed through a duct into the digestive system. Some of the more valuable enzymes are: *Pancreatic lipase* which acts upon fats that have already been emulsified by bile and transforms them into fatty acids and glycerol, a fatty substance. You need fatty acids to provide a concentrated source of calories for energy. Fats also aid in storage and assimilation of vitamins A, D, E, and K. You also need fats to aid in certain processes of metabolism and cell structure. Fatty acids help build the structure of your brain and nerve tissues. Fats also maintain your skin health—serving as a cushion for your body against blows and bruises; fats also make this body covering (your skin) almost impervious to water. So, you do need fat! Pancreatic lipase is the *only* effective fat-splitting enzyme. Without it, fats cannot be properly digested or absorbed by your body.

Trypsin is a pancreatic enzyme which acts upon proteins; particularly those which have not been partially digested in the stomach. Trypsin also acts upon the foods that have been partially treated by pepsin, breaking down these substances into *peptides*.

Amylase, as a pancreatic enzyme, serves to split uncooked as well as cooked starch into maltose—a substance that can then be utilized by your body.

Stomach enzymes.

The mucous membrane of your stomach contains glands which secrete valuable digestive enzymes, as we have seen; also, the stomach issues the most valuable (if any one enzyme must be singled out) "fountain" known as hydrochloric acid. Edith E. Sproul, M.D., lauds the power of this enzyme by stating, "Hydrochloric acid serves many useful purposes. Its secretion is related to the acid-base balance of the body and is accordingly linked to the kidneys' elimination of an acid urine and to the lungs' blowing off of alkaline gas."

"The acid nature of the gastric juice is responsible for the destruction of bacteria in the stomach, so that the foodstuffs enter the intestine virtually in a sterile state. Hydrochloric acid also liberates iron from the food and converts it to the form required for absorption. This is an important contribution to red-blood-cell maintenance."

Hydrochloric acid works mainly upon tough foods such as fibrous meats, vegetables, poultries, etc. If you are deficient in this valuable stomach enzyme, you run the risk of improperly digested foods as well as loss of extraction of valuable vitamins and proteins which your body must have.

The big job enzymes do.

In brief, we see that enzymes perform just about every function needed to build and maintain life and health. Enzymes transform foods into muscle, nerve, bone and gland. Enzymes help to store excess foods in your muscles or liver for future use. Enzymes also aid in the formation of urea which is passed off in urine; enzymes aid in the passing of carbon dioxide from your lungs.

You have an enzyme that helps build phosphorus, the mineral, into nerve and bone. Another enzyme helps fix iron in your blood cells. Those who suffer from poor blood or conditions of anemia, will be interested to know that not only must iron-rich foods be included daily with their meals, but they must have a strong enzymatic system which is needed to metabolize the iron for blood utilization.

Enzymes aid in blood coagulation and thereby stop bleeding. Enzymes serve to decompose poisonous hydrogen peroxide and liberate healthful oxygen. Enzymes also promote oxidation—uniting oxygen with other substances to aid in the process of breathing. Enzymes attack toxic substances in the tissues and blood, changing them into uric acid and urea so they may be passed off and eliminated from your body.

You have already seen how enzymes transform proteins, carbohydrates, fats, vitamins and other nutrients into usable substances.

The Columbus, Ohio, *Health Bulletin,* in urging greater understanding about enzymes, declares that the "amino acids originating in the muscles of the steer which becomes our steak are transported by the bloodstream for the construction and repair of our bodies. Along the way these acids are captured by enzymes and converted into muscle, a radically different substance from the original. We chew a bite of bread and it becomes sweet. The starch in the bread, which is indigestible, has been changed into sugar by enzyme action. *Invertase,* an intestinal enzyme, can break down a million times its own weight in sugar and be ready for more."

The power of enzymes is seen in every bit of food that we eat; most food is not easily or instantly digestible (that is, it cannot be immediately absorbed into the bloodstream until it is processed by enzymatic action). As we have seen, when you eat a steak, a celery stalk, you are not eating fuel, energy, or building blocks of the body. Eat raisins, but you do not eat iron as such—you eat foods containing these nutrients which enzymes must extract for utilization by the body.

If you blink your eyelid, chase after a commuter train or bus, cogitate upon a financial schedule, study a recipe, or keep your heart beating— enzymes make these processes possible for you!

Basic enzymes sources.

Where does your body obtain the material from which the enzyme is manufactured? There are three sources—and, it must be added, the *only* sources: *air*, *water*, and *food*. Certain substances in air and water aid in the formation of materials which then create enzymes. (See chapter on proper breathing instructions to create a healthy enzyme environment.)

Raw foods are the prime sources of enzymes. Aside from the changes that take place in cooked foods, enzymes are destroyed under certain conditions of heat.

Raw fruits, vegetables, eggs, unpasteurized milk, meats, fish, etc., are all rich in enzymes—when they are uncooked and unprocessed. Of course, it is unpalatable to eat raw eggs, just as it is disagreeable or distasteful to chew on raw meats—not to mention the health hazard since uncooked meats have been found to contain parasites which may lead to disease. Because enzymes are virtually annihilated by processes of boiling, baking, frying, stewing, roasting, etc., this means you must select those foods which can be eaten raw. These include fruits, vegetables and certain wheat products.

Raw, or *unprocessed wheat kernels* are good sources of the enzymes, amylase and protease. While grinding wheat into flour will not destroy these enzymes, to bake the flour into bread requires heat and therefore destroys these vital substances. Eat raw or uncooked wheat kernels for a good enzyme source.

Health food stores and special diet shoppes carry *raw wheat germ* in handy one-pound packages, or in smaller containers that may be tucked in your purse or coat pocket.

Certified raw cow's milk, on sale at many health food stores, would contain valuable enzymes. When milk has been pasteurized, it has been put through a special heating process which destroyed the valuable enzyme content.

Certified raw goat's milk is another hitherto undiscovered source of highly concentrated enzymes. Here again, health food stores are outlets for this natural food.

Fresh fruits and vegetables are valuable sources of enzymes. When canned, the processing destroys enzymes in either the edibles themselves or in their extracted juices. One of the best sources of enzymes is found in natural raw fruit and vegetable juices—which are extracted on your home juicer. If you are able to afford it, obtain such an electric juicer (up to a dozen different varieties and styles are to be seen at your local health food store) and squeeze your own liquids—rich in enzymes. *Drink the juice immediately* after it is squeezed.

Wherever possible, eat fresh raw foods.

Before you start a meal, eat a piece of raw fruit or raw vegetable—your dessert should consist of a raw fruit or raw vegetable salad. Such foods as carrots, cabbage, Swiss chard, broccoli, onions, celery, turnips, can be eaten raw! If you find them a bit strange to your palate, make a slow adjustment by drinking their juices, made in your own home, if possible.

In preparing raw vegetable dishes, use a little imagination. Try a grater, a foodmill, a chopping board, a blender. Add zest to a plain raw vegetable plate by using a salad dressing—made of equal parts of fresh apple cider vinegar and liquid corn oil. These last named foods are sold in all health food stores throughout the country. As "fixings," use dates, raisins, nuts.

It is believed that fruits and vegetables that have been organically-raised have a better source of enzymes. Time and again, scientists have found that when test animals are fed organic foods, they have greater resistance to diseases. Sir Albert Howard, in *The Agricultural Testament*, tells how his cattle were fed organic food—while on an adjoining farm, another herd of cattle was fed foods treated with chemical fertilizers and insecticide.

Sir Albert's cattle had no signs of disease—even though they rubbed noses with the neighboring cattle which had hoof and mouth disease! Yet, the organically fed cattle remained absolutely free from the epidemic malady. It is possible that the powerful enzyme treasure in the organic foods helped build resistance to disease. (Chemical fertilizers and insecticides have been seen to reduce and also destroy enzyme content in foods.)

So, the next source of enzymes would be in organically-raised foods. If you have a small plot of ground where you can grow your own, that is fine. If you do not, seek out farmers who sell organically-raised produce. You may find some in your area by writing to your state's Secretary of Agriculture. The little bit of extra effort you extend will be well worth the reward of increased good health that is possible from the supply of enzymes in natural foods.

CHAPTER 1 AT A GLANCE.

1. All bodily processes, mental and physical, healing and maintenance are ruled by enzymes, those substances which have the power to rejuvenate an entire being.

2. Enzyme juices are secreted by your digestive, salivary, intestinal, pancreatic, stomach systems.

3. Your body must have fresh air, pure water, nourishing foods from which they can make enzymes.

4. Raw foods are your greatest source of enzymes. Yes, you can eat your way to enzyme power by stepping up your raw food intake.

2

MAKE DIGESTION WORK
FOR YOUR HEALTH—NOT AGAINST IT

Why do some folks eat copious amounts of nourishing foods, yet fall vulnerable to ailments, premature aging, chronic fatigue? Why do other folks eat moderately of less nutritive foods, yet show more health than others? Apparently, the answer lies in the proper digestion and absorption of foods that you eat. No matter how high the quality of your foods, no matter how carefully prepared, you still run the risk of declining health if your food does not feed you.

That's right—your food must feed your system all of its nutrients and this is accomplished by proper digestion via enzymes. If you lack digestive enzymes, then digestion works against your health by failing to extract vital life-giving vitamins, minerals, proteins, amino acids, essential fats, etc.

Basis of enzyme power.

Good enzyme power has its foundation in good chewing—which, in turn, aids and dominates the entire digestive system. Just how do enzymes work to release the life-building elements that are absorbed into your bloodstream to be carried to all body parts for your mental and physical nourishment?

Good digestion begins in your mouth! This means that you must chew and chew well if you expect enzymes to do a good job. "The first enzyme in the digestive tract is found in the saliva of the mouth and is a starch splitter called ptyalin," explains Catharyn Elwood, in *Feel Like a Million.* "To be completely digested, all starches should be thoroughly chewed and mixed with ptyalin. It is easy to experience the enzyme action in your mouth by chewing a dry piece of whole-grain bread, toast, or any other starchy food. The flavor will soon change to a distinct sweetness as you continue to chew.

"The increased sweet flavor is the changing (digesting) of starches to simple sugars. If they do not become simple sugars, they cannot be absorbed and converted to energy. . . . If starches are bolted or washed down with liquids, the food passes by so quickly it is not mixed with ptyalin. An added burden falls on the starch-digesting enzymes in the duodenum and small intestine, and often the starch digestion is incomplete. The undigested foods pass from the body and are lost. Eating too fast can indirectly contribute to fatigue and weariness."

Dr. Elbert Tokay, in *The Human Body*, declares that chewing is most valuable for enzymatic power. "The teeth and the muscular tongue are important in chewing and the latter starts the food on its journey through the digestive tract. Most of the saliva (containing enzymes) that pours into the mouth is secreted by three pairs of salivary glands. The gland cells secrete saliva into small ducts which unite to form larger ones and finally one or two large ducts which carry the fluid into the oral cavity."

How to chew for best enzyme action.

Chewing enables mouth enzymes to get at the food, preparing it for the journey which will be described in a few more paragraphs. Here is a 3-step chewing plan designed to create a favorable enzyme environment:

1. All fruits, vegetables and sugar-starch items must be thoroughly chewed. Starch and carbohydrate foods must have an alkaline base for thorough enzymatic action. Mouth saliva is a good alkali source. Your salivary glands will secrete this essential alkali base—only by chewing! If you bolt down foods and care little about proper mastication, then solids will go into your digestive tract, unprepared by alkali, rendering them less digestible and less nutritious.

2. Thorough chewing increases the enzymes in your digestive tract—causing a flow, even before such foods have been swallowed! The more abundant your enzyme flow, the more digestible will the foods be rendered; at the same time, chewing causes an involuntary secretion of pancreatic fluids, bile and intestinal secretions—all of which are needed to attack foods and release their precious life and health building nutrients.

3. All protein foods, including meat, fish, poultry, eggs, cheese, beans, nuts, etc. are also in need of chewing but they should be made comfortably soft and then swallowed. The real work of digesting proteins is done in the stomach and hydrochloric acid is far more powerful in pulverizing meat fibers than your set of teeth. Chew these foods carefully, of course, and swallow them when comfortable to do so, but you need not liquefy them as you must do with starch foods as well as fibrous fruits and vegetables.

Now, let's look at your digestive system and see how it works with enzymes to provide instant health. Your tongue is a power of muscular energy and force. Your tongue pushes foods from one side of your mouth to the other, while you chew. The tongue surface is a series of small sense organs or taste buds which are scattered among the little papillae—small surface bumps. You possess four fundamental taste reactions—sweet, sour, salty and bitter. These taste reactions are influenced by your nasal olfactory organs. For example, suppose you have a heavy cold or stuffed nose. You immediately lose some powers of your taste buds.

Digestive enzymes have a reaction to the sight, taste, smell and thought of foods. Foods which you do not like will inhibit enzymatic secretions. You reject them! A coated tongue in an illness will also spoil gustatory reflexes, spoil your appetite and diminish enzymatic flow. Worry, emotional or mental strain, also has a bad influence on enzymes. We all know how fear, anger, tensions, interfere with digestion. Never eat when emotionally or physically upset.

Now, as your teeth and tongue pound food into a bolus, a pulpy mass, your stomach is preparing itself by issuing forth enzymes to help digest these foods, and extract their nutrients. The enzymatic flow may be hampered if you are tense or nervous or eating under uncomfortable circumstances, so even though you chew properly, you still run the risk of an upset or heavy stomach.

When food is completely chewed into its mass, or bolus, it is now ready to be swallowed so it can enter your stomach. To enable you to swallow, your cheeks, tongue, roof of the mouth muscles all cooperate to form a sort of chute. Tongue power against the strong palate of your mouth roof now pushes this mass into your pharynx—a passageway which transports food into your stomach; the pharynx also serves the purpose of acting as a tunnel, bringing air into your lungs.

If you talk while you chew and eat, you may suddenly start choking and sputter out, "It must have gone down the wrong way."

Actually, what happened is that talking caused both air and food to enter your pharynx at the same time. Proper digestion requires that these two functions be kept apart. So—don't talk while you chew and swallow!

Your soft palate raises up as food travels along the back of your mouth. During the swallowing action, the opening to your nasal cavities and windpipe shut tight automatically. Should food or water enter your larynx while you are swallowing, you start to cough, sputter and choke.

How peristalsis acts.

A wave of muscular action performed by the non-striated or involuntary muscles, shoves the food bolus into your esophagus (a 9-inch stretch of gullet or food tube which conveys swallowed foods and liquids from the back of your throat down into your stomach), which reaches from your chest cavity into your stomach.

Your stomach, shaped like the letter "J," about ten inches long, is made of thick, muscular walls which have solid layers of circular, longitudinal and oblique muscle fibers. Stomach muscles are needed to breakup food into still smaller particles.

Improperly chewed foods require more stomach muscle action, causing a strain on the digestive system because your stomach must work harder to accomplish what your jaw muscles should have done. This bears out the truth that if you gulp down your foods without vital chewing, you run the risk of cramps and stomach pains because of the vigorous actions of these muscle fibers.

Your complete digestive system is lined with a mucous membrane containing small glands. In your stomach, located behind these tiny mucous glands are the still smaller gastric glands which issue the vital enzyme juices. It is believed that your stomach lining possesses at least 35 million of these smaller gastric glands.

Your digestive processes.

Among the enzymes secreted by these glands is hydrochloric acid—so powerful that if your stomach did not have its protective thick mucous covering, this enzymatic acid could actually dissolve the stomach itself. Hydrochloric acid is needed to attack meats and heavy protein foods, extracting its valuable nutrients.

Rennin, the other enzyme, now comes out to curdle cheese products.

Lipase works to split up such fat containing foods as egg yolk, creams. Hydrochloric acid will join with proteins to help create pepsin which splits other particles into still smaller units, then digesting the milk curds created by rennin.

These miracle workers of Nature, these enzymes, are so powerful, they could even digest the entire stomach! But Nature has seen fit to protect you by causing the stomach lining to secrete small amounts of ammonia from tiny cells. This natural ammonia substance serves the purpose of neutralizing and counteracting the power of hydrochloric acid against the stomach lining.

The bolus food mass is given this enzymatic treatment from three to five hours. About three pints of enzymatic fluid are secreted during a single twenty-four hour period. About one and a half pints are needed to digest a hearty meal. A healthy enzymatic system should produce about two-thirds of an ounce of hydrochloric acid in twenty-four hours. Of course, the amount varies with the foods you eat. Meats need stronger enzymatic actions.

Gradually, the bolus food is changed into *chyme*—a semi-fluid. Creamy, it is ready to pass into your duodenum—the first portion of your small intestine. Here, more enzymes attack the chyme. Enzymes from your pancreas, bile and liver are sent forth to combine with the chyme; at the same time, tiny glands along the walls of your intestines secrete more juices to act upon the chyme.

Meanwhile, enzymes in the pancreatic fluid will be breaking protein into amino acids to provide the building blocks for cellular growth; large sugar molecules are changed into simple and digestible sugars. Fats are transformed into fatty acids to prepare them for absorption into the bloodstream or lymph vessels.

(Lymph is a fluid that is the color of straw; it is the intermediary between the blood and tissues since blood is sealed within the vessels and does not come into direct contact with tissue cells.)

Just how fast are these substances handled? Among solid foods, carbohydrates are utilized most speedily by the stomach's muscular contractions and digestive juices. Proteins need more time. Some fats will hinder the digestive process for other foods in the stomach since fats have a way of slowing down the secretion of gastric juices. A fat-heavy meal may need longer digestive and assimilation time.

Peristalsis continues to push the chyme into your small intestine—a 20-foot length of tube, four times as long as the large intestine.

During this process, the pancreas has been working to secrete insulin, directly absorbed into the bloodstream. An ailing pancreas or one that does not secrete sufficient insulin because of a physiological disorder, means that the condition of diabetes occurs.

Your digestive system needs insulin to dissolve and metabolize sugars. Without insulin, too much sugar remains, enters into the bloodstream and causes possible illness. So, it becomes apparent that you need every single enzyme to perform life processes!

Now your enzymes and digestive fluids have prepared the foods to be absorbed and assimilated. The small intestine's interior is covered with endless fingerlike projections called *villi*. Imbedded in the villi are still tinier branches of two sets of vessels. The first, the *lacteals*, belong to the lymphatic tubes. The second, the *capillaries*, belong to the blood vessels.

The small intestine has millions of villi. In constant movement, swinging back and forth, they are always active. When seen on a large screen, the villi are similar to a huge wheat field, swaying in the breeze. These villi are endowed with astonishing powers of absorption. The lacteals of the lymphatic tubes extend to the thoracic duct, the main reservoir which reaches upward before the spinal column, then enters into one of the main blood vessels.

All properly digested food is sucked by the villi into this thoracic duct; the extracted nutrients now enter your bloodstream to nourish your entire body, your tissues, cells, from head to toe.

The capillaries of your blood vessels now converge into a still larger vessel called the portal vein which extends into your liver. Through this vein travel whatever nutrients the capillaries have absorbed. Once they reach your liver and bloodstream, these absorbed food elements undergo still further modification while being prepared for assimilation by body cells.

(By the way, sugars and amino acids enter blood capillaries; fatty substances usually go into your lymph vessels.)

Chyme, and what's left of it.

By this time, the chyme has reached the end of the small intestine. What is left of the chyme that started out as food put into the mouth?

Just water and waste substances. This chyme travels along into the large intestine, about five feet long. It (the large intestine) looks like a picture frame, reaches up, across, then back down through the stomach cavity. The large intestine has no villi, since none are needed. By the time the chyme reaches the large intestine, all valuable nutrients should have been extracted by the enzymes.

The chyme is now a mass of water and indigestible residue. The walls of the large intestine absorb much of this water to be passed off as urine. Semi-solid waste substances are evacuated during a bowel movement.

What happens to nutrients?

Now, what has been happening to the valuable nutrients extracted by your enzymes? Your blood vessels are active in carrying many of these elements to the liver—the largest organ in the human food factory, as the digestive system is called.

Your liver has the job of absorbing fat products, creating fuel, body heat and energy—energy to lift your finger, raise an eyelid, life a heavy bundle—energy to think!

Your liver stores vitamins and minerals; your liver also processes iron to be used by the bloodstream. The liver also dispatches such nutrients as sugars to the cellular body tissues where these same sugars are metabolized (burned) to give caloric action.

All in all, the complete digestive process which began when you started chewing food, until your bloodstream carried vital nutrients to nourish your body, a time period of 24 to 30 hours has elapsed. Of course, this all depends upon the foods you've eaten, their quality and quantity, and also your physical and mental emotions.

Nervous stomach.

Some years ago, the famed Dr. W. B. Cannon of Harvard Medical School, noted that *four hours* after a dog was made nervous by seeing a cat, he still had impaired digestive functions.

If you complain of a nervous or frequently upset stomach, you must cultivate the art of relaxation so that when you do eat, your digestive system will work favorably. Fear, worry, anger, tension, emotion, all cause an inharmonious digestive environment; these conditions tend to suppress the flow of enzymes and also inhibit the muscular (peristalsis) activity of the alimentary tract.

A stomach that issues warning signals including gas, heartburn, nausea, heaviness, cramps, fever, headaches is begging for rest! You avoid eating under such circumstances.

Ever notice how your appetite increases with an automatic saliva flow when you see food you know you will enjoy? This can work in reverse. Strong emotional upsets have an inhibiting effect; not only will it cause a dry mouth but insufficient saliva will slow down the muscular movements of the alimentary tract. The complete digestive function is impaired, creating what is called a "nervous stomach," yet is actually an enzyme-deficient system!

Tense nerves cause food to be metabolized at an unnatural rate; lactic acid accumulates to be carried to the liver where it is converted into body starch to maintain the body tension. Some lactic acid remains inside the liver since not all of it can be converted so rapidly to meet the body demands. This causes irritation, stomach pains, poor digestion and overall ill health. Learn to avoid eating if under severe tensions.

How to cultivate a healthy enzyme environment.

Thought control and self-hypnotism are needed to spark enzymatic function before you even touch your food. This is accomplished by understanding the old proverb, "One picture is worth a thousand words."

With this in mind, you set up a plan to visualize yourself as being in a tranquil, calm and placid state of mind. Picture yourself walking to the dinner table, sitting down, chatting with friends, putting a snowy napkin on your lap, then awaiting the first course of the meal that contains all of your favorite foods. You need to keep fortifying this image as a means of self-induced relaxation.

Hear what David Seabury has to say in his book *The Art of Living without Tension* as a first step in creating a favorable enzyme environment.

1. "Sit back and close your eyes for a few moments. As you do so, imagine that a motion picture is before you in which you are playing the lead. Let your emotions rise and go into the picture. It should be a portrayal that concerns some special desire of yours, a drama about some longing you greatly wish to have fulfilled. Picture it happening with you yourself intensely in it: that is, heroically conquering the difficulties in the way. . . . Live this emotion every evening. . . . Go back every night to the same picture. Recreate it. Improve it. Add conclusive touches

here, shorten it in places that move too slowly. Work at it as faithfully as a scenario writer would in order to give you a dynamic picture." Now let us go on with successive steps:

2. Now you are calming your whole mental and nervous system by transporting your whole being in a place of calm tranquility. You shut out all negative thoughts. You banish worry. You prepare yourself for a nourishing meal. Next, begin your meal with either a fruit, cup of vegetable soup, meat broth or bouillon. Why? Because juices from fresh fruits and vegetables, as well as meat extract or bouillon, (especially when made by simmering together meats and vegetables) are prime sources of *extractive substances*. When these substances reach your stomach before the main foods, they influence specific cells in your lower stomach to issue a hormone called *secretin*. This hormone passes right into the bloodstream, then is carried to all body organs, including the glands in the stomach's upper section. Here, the secretin hormone stimulates the digestive glands into action, urging them to issue sufficient amounts of powerful hydrochloric acid needed to digest protein foods. Other enzymes are also sparked by secretin—which depends upon the extractive substances found in the aforementioned foods: fruit, vegetable soup, meat broth of bouillon. Did you know that this digestive secret has been known by the Europeans and Orientals for centuries on end? It is rare that you sit down at a European table and jump right into the main course. No. First you must prepare the stomach for its forthcoming digestive job—and this is done by first sipping fruit or vegetable juice or a meat broth. We Americans have lots to learn about creating an enzyme environment.

3. The manner in which you chew influences tension! Dr. G. I. Freeman, writing in *Psychological Review*, reports on experiments that chewing slowly and carefully is beneficial to the entire digestive system . . . and also performs a relaxation effect upon your nerves. If chewing is done strenuously, then body tension is increased. It is suggested that you take your time in chewing, at a rate that is comfortable and satisfying to you. Just as it is harmful to the enzymes to bolt down partially chewed food, it is just as inharmonious to chew at a fast and impatient state, and swallow in that frame of mind.

4. In times of illness, avoid complete meals as you would ordinarily eat when perfectly healthy. A coated tongue is the first symptom of an illness; the fact is that when your tongue is coated, the taste buds are

dulled or completely nullified; this prevents you from normally appreci-
ating the food flavors, prevents the establishment of gustatory reflexes
and hinders the secretion of sufficient enzymes. When you are able to
taste your food, then you enjoy the flavors, lingering longer over each
mouthful and are not tempted to hurry through the meal with a gorging
process. A keen hunger and the ability to heartily enjoy food is an indi-
cation that there will be produced a full supply of the requisite enzymatic
juices. The more you enjoy your food, the more you delight in chewing
to extract every mouthful before swallowing; this helps in an increased
flow of all gastric enzyme juices.

5. Eliminate condiments such as harsh spices, salt, catsup, etc. They
not only irritate the enzyme forming organs, but impair their functioning
powers. They blunt the sensibilities of the gustatory nerves and thereby
diminish your enjoyment of foods. Condiments also cause increases in
the flow of certain digestive juices which are more mucous than enzyme-
rich. Hot spices irritate the stomach and cause it to dispose of its foods
as rapidly as possible, preventing the completion of the full process of
chyme. Condiments may also cause a premature emptying of the stomach
before valuable nutrients have had a chance to be extracted.

The irritation caused by mustard, pepper, horseradish, vinegar, etc.,
may produce blisters upon the skin when administered in a pure state.
When taken internally, they exert their irritating effect upon the more
delicate membranes of the digestive system, excite the stomach to increased action in certain respects, but lessen the secretion of gastric juices
and later decrease activity of the entire digestive system. These substances act upon the digestive organs as a lash and the spasms they
induce are harmful to digestive processes.

To take the place of harsh spices, seek out such substitutes as salt
substitute powders, anise seeds, celery and caraway seed, sage, and the
popular "vegetable salts." These are sold in special diet shoppes throughout the country, as well as in the popular health food stores.

6. Avoid drinking while you eat. This habit leads to the bolting of
food. If food is washed down instead of being properly masticated, it
reaches your digestive system in an unprepared manner. Many foods are
dry and require much insalivation before they can be swallowed. The
purpose is for you to chew and let the enzymes in your mouth glands do
part of the job of extracting valuable nutrients from such dry foods. To
wash down with a lot of drink will prevent the completion of this first

and vital step in digestion. If you avoid harsh condiments in your foods, you will be less thirsty. Also, eating green and succulent vegetables with your meals will give you sufficient natural water to lubricate the foods that are dry. Drinking with meals has a way of creating a ravenous appetite and leads to overeating.

Finally, outside noise while eating is disturbing to digestion. Salivary and gastric secretions "freeze up" under tension coming from noisy crowds, excitement and emotional stress. Enzymes delight in quiet, cheerful surroundings—so eat when you are in the same frame of mind!

When you're happily content—so are your enzymes.

HIGHLIGHTS OF CHAPTER 2.

1. Diet decides whether you are to be fit as a fiddle or victim of constant ailments.

2. Sound digestion starts in your mouth where salivary enzymes begin extracting body-building nutrients from foods that you chew.

3. Enzymatic power requires you to chew thoroughly all foods—especially raw fruits and raw vegetables.

4. Chew thoroughly all protein foods to ease the work of your digestive enzymes.

5. As you chew, swallow and digest foods, your entire enzymatic system is at work assimilating these substances and extracting life-giving nutrients to nourish you from head to toe.

6. Enzymes work best when you are at peace. Before you eat, get yourself composed and relaxed. If you cannot eat in a tranquil surrounding, don't eat at all and wait until you can have peace and quiet.

7. Avoid harsh spices and condiments since they are at odds with enzymes.

8. Avoid drinking while you are eating. Don't overload your enzymes with too much liquids while you eat. If this happens, *you can drown your enzymes.*

3

ENZYMES CAN HELP CONQUER
DISEASE

"Enzymes produce the miraculous changes taking place in the budding of trees and shrubs and the production of fruit. All the living things that grow on this planet owe their being to enzymes," declares Dr. Alice Chase, author of *Nutrition for Health*.

"More complex and less familiar are the glandular products," continues the doctor. "These substances regulate the growth, health, well-being, happiness or misery of the individual. An example of one of these enzymes is thyroxin, the basic product of the thyroid glands. The hormones of the sex glands illustrate another type of enzymes. There are enzymes in various tissues that promote their nutritive functions. In disease, the body turns either fat or thin through the action or lack of action of enzymes."

Disease-fighting powers of enzymes.

Dr. Chase then points out the disease-fighting powers of enzymes. "Among other enzymes that play a role in disease are those that produce *autolysis*. For example, a tubercular lung may disintegrate through the action of enzymes as well as of TB organisms. Cancerous tissue often disintegrates because of characteristic pathogenic enzymes. An ulcer of the stomach or elsewhere may be produced by pathogenic enzymes." (By pathogenic is meant diseased. Therefore, an enzyme that is diseased is regarded pathogenic.)

"There are other enzymes," continues Dr. Chase, "that regulate the health of the body, and also other enzymes that disintegrate tissue in the course of disease. In the late stages of cancer, TB or diabetes, there is a wasting of the tissues. Diabetic gangrene in terminal structures may be caused in a measure by enzyme action. Spontaneous fracture of bones is caused by a lack of minerals and vitamins in the bone structure; but certain types of enzymes may be responsible for this form of degeneration."

Modern therapy is aware of the healing powers of enzymes and is successfully extracting them to be used in the manner of antibiotics and other medications.

A pioneer researcher was Dr. James B. Sumner of Cornell University. Back in 1926, Dr. Sumner isolated one enzyme from an extract of jack beans—a legume that is a prime source of numerous enzymes. Most vegetable stores sell jack beans and customers are unaware of its potential as an enzyme treasure. Dr. Sumner extracted *urease*—the enzyme that breaks down urea, a substance formed within the liver and delivered to your kidneys by way of the bloodstream. It is a normal and healthful function for you to excrete about an ounce of urea daily. To enable this urea to be in a form that can be passed off, it must be broken down by *urease*—the enzyme found in jack beans. Without this enzyme, it is possible for a toxic or poisonous condition to take place in the bloodstream. It is this enzyme—urease—which helps to assimilate internal urea.

Solvent powers of enzymes.

Four years later, Dr. John Howard Northrop of Rockefeller Institute, was able to isolate pepsin—heretofore believed to be important in digestion of protein foods. The discovery that isolated pepsin could be given to those with weak digestive powers or suffering from protein deficiencies, opened a new world of attempts to isolate other enzymes.

Nearly all current isolated enzymes are lytic—that is, they dissolve something, splitting molecules into fragments.

A leading health magazine, reporting on enzymes, declares, "This (lytic) action can be very comforting, in liquefying pus and promoting drainage, and in softening sticky mucous plugs to help patients with bronchial asthma, emphysema, and similar conditions to breathe more freely. It can even be lifesaving.

We are all familiar with *papain*, the enzyme of the fruit, papaya, which works to tenderize meats. It is similar to pepsin in digesting protein, aiding in transformation to amino acids that are taken up by your bloodstream and sent to nourish your whole body.

Another extracted enzyme is cellase, which has the unique power to digest so-called indigestible cellulose, the tough and fibrous "stuff" of such coarse foods as celery, carrots, root vegetables, etc.

At one time, a major surgical procedure was to clean away from wounds such deterrents as dead tissues, clots, pus, contaminants. Healing could not otherwise be promoted without such cleansing action. Today, enzymes are used to flush away infection-breeding debris with mild gentleness. We all remember back in the days of World War I when it was the vogue to use maggots for wound cleansing; undoubtedly, an enzymatic action was created that helped perform this cleaning and enable healing to take place.

A blood enzyme, fibronolysin, is needed to dissolve strong fibrinous deposits and clots. The bulk of pus consists of cellular material which can cause serious infections. Here, ribonuclease enzymes are used to dissolve such cellular materials. These are prepared from beef pancreas and sold as *Dornase*, in a commercial compound, available on prescription from your doctor.

Two other pancreatic enzymes—*trypsin* and *chymotrypsin*, are used to counteract inflammation and bring down swelling. In such ailments as bruises, fractures, sprained ankles, back injuries, these enzymes are valuable.

Those who are allergic to penicillin will be glad to know that we have an enzyme called penicillinase—it eases the allergic symptoms to this antibiotic and helps you build tolerance to it.

Armour has developed *chymotrypsin* in a special form, to be used during such ailments as cataract surgery. In this operation, a surgeon must remove the diseased lens—if he has this enzyme helper, he may not have to cut the tiny ligaments that hold the lens in place. Instead, the surgeon injects chymotrypsin and this enzyme selectively dissolves the anchor-fibers; the lens floats loose.

Procollagenase, another isolated enzyme, is being used to "erase" disfiguring scars and intestinal adhesions after surgery. It has the ability to dissolve material from which tough fibers develop.

Troubled with leg clots? It has been seen that fibrinolysin will dissolve them, depending upon how soon given after formation of the clot. It is already believed that enzymes can be administered to dissolve coronary clots or cerebral arteries related to heart attacks or strokes.

When disease strikes the body, it has been noted that enzymes are

forced into unnatural activities. The Columbus, Ohio *Health Bulletin*
declares that "many medical and research authorities are of the opinion
most diseases originate from lost or deficient enzymes.

"Faulty enzyme behavior is at the root of leukemia and other cancers,
some believe, with the possibility that synthetic enzymes of positive or
negative action eventually may be utilized to hinder overactive enzymes
or to assist those which have become ineffective to conquer many diseases
with which man is saddled."

Other diseases which include mental disturbance, are also traced to
malfunctioning of enzymes.

Dr. Felix Wroblewski, of the Sloan Kettering Institute for Cancer Research in New York City, set about to research the manner in which
certain diseases change the composition and the amounts of enzymes in
the human body. It is believed that when the body has either an under-
production or an overproduction of enzymes, disease is able to strike.
By controlling enzymes, a cure can be possible.

As an example, Dr. Wroblewski found that after a patient had a heart
attack, within the first 24 hours of the seizure, the blood enzyme
quantity almost tripled!

Infectious hepatitis, a disease striking the liver, causes an abnormal
increase in the enzyme, glutamic pyruvic transaminase—as much as ten
times in the bloodstream. When the doctor takes blood tests at different
intervals following the first diagnosis, he can actually tell the rate at
which the disease is developing, and know what to do, basically by
following the enzyme count!

Trypsin, an enzyme secreted by the panceras, may help overcome
problems of circulatory ailments. The value of trypsin in the treatment
of thrombophlebitis (leg blood clots) was confirmed at a Conference
held at the New York Academy of Sciences. Furthermore, New York's
Bellevue Hospital reports about giving trypsin by injection to 82 cases
of thrombophlebitis in what is known as a double-blind test. That is,
under a prearranged plan, neither doctors nor patients know what is
being used. Some patients receive the enzyme; others are given a placebo,
or inactive substance. A special coding system is put into use by a third
party who is the only one aware of the identity of those receiving the
true or false enzyme.

Stating it simply, the trypsin enzyme is able to dissolve protein—and
clots are composed largely of protein substances.

At an annual meeting of the American Chemical Society in Atlantic City, New Jersey, it was stated that *elastase,* an enzyme secreted by the pancreas, was successful in dissolving cholesterol deposits in the arteries of test animals. Research is to be conducted with humans to see if excess cholesterol, the forerunner of heart attacks and arteriosclerosis (hardening of the arteries) can be reduced by enzymes.

Dr. John C. Houck, of the Childrens Hospital Research Foundation in Washington, D.C., as reported to *Science News Letter,* found that procollagenase, another enzyme, is actually able to prevent or melt ugly scars and render them invisible. How? What is a disfiguring scar? It is a substance of collagen, a fibrous protein. The procollagenase enzyme is able to dissolve this form of collagen from which scars develop.

A spokesman for the Amateur Boxing Association of London, Dr. J. L. Blonstein, said that he treated boxing injuries by combining two enzymes —streptokinase and streptodornase. When absorbed into the bloodstream, they dissolved blood clots and broken down tissues. "We have been able to get a 50% reduction of bruises and abrasions and a 15% reduction of blood swellings; and cuts, which would have normally taken four weeks to heal, have done so in one or two weeks," said Dr. Blonstein to the London *Times.*

Those worried about menstrual pain may welcome a report made by Dr. Ralph Heinicke of the University of Minnesota. The doctor found that *bromelin,* an enzyme found in pineapple stems, was given to women who suffered from monthly tension pains, and also to those who had difficult childbirths. It was found that the menstrual pains were eased and that this pineapple extract relaxed the female pelvic muscles to facilitate the birth of a child.

Philip J. Pollack, M.D., writing in *Current Therapeutic Research,* tells how he treated 49 patients—who had such conditions as hemorrhoids, heart disease, skin allergies, bed sores—with papain, the enzyme taken from the ordinary papaya. One such typical case was that of a 52-year old woman. She suffered from extremely painful external hemorrhoids. She had been treated with bed rest, sitzbaths, rectal suppositories—with no response.

Dr. Pollack gave her one papain tablet every 4 hours. Within 48 hours, the patient showed "marked improvement." The painful swelling was reduced. Three days later, her improvement was regarded complete. She needed *no* surgery. All this was brought about by an enzyme. How did it work? It reduced inflammation by dispersing fibrin deposits—this, in turn, worked to drain the fluids that caused the swelling; at the same time, the useless dead cells were removed. Now that these obstructions were removed, the body was able to go about its task of repairing tissues easily and speedily.

Slipped disk therapy.

A slipped disk is a condition in which one of the intervertebral disks (cartilages found between the vertebrae of the spinal column) usually in the lower back, has slipped out of place. You need these disks because they function as small cushions to absorb the shock on the spine from ordinary movements. Sometimes, the disks may break (rupture or herniate) and cause severe low-back pain that extends downward to the back of the leg and into the heel. A person with a slipped or ruptured disk feels severe pain just when bending over. Surgery is usually regarded the best way to correct such a condition.

But now, it is found that enzymes may even overcome this condition, freeing the patient from surgery. Papain has a dissolving power. With this knowledge, a Chicago orthopedic surgeon, Dr. Lyman Smith, used papain to inject into the nucleus of the disk of a test animal. Papain went immediately to work to dissolve the nucleus, quickly eliminate the pressure and the pain. Says Dr. Smith, "We have high hopes that this same process may someday be used on humans. . . . It won't be today or tomorrow. But we do think the enzyme (papain) has excellent possibilities in this field."

Mental retardation.

Mental retardation is another condition which doctors find may be traced to a lack or deficiency of an enzyme. Called *histidase,* a shortage impairs memory of youngsters. Dr. G. Donald Whedon, director of the National Institute of Arthritis and Metabolic Diseases, has this to say, "Most of us are able to speak tolerably well and continue a logical conversation because we hear what we say over a sufficient period of time. . . . But these (retarded) individuals apparently are unable to hear their voice for more than a split second, and so cannot put together sufficient verbiage to make appropriate statements."

It is felt that histidase has an influence upon the memory section of the brain.

Palsy conditions.

Another enzyme breakthrough is indicated with the condition of Parkinson's disease (shaking palsy). A degenerative disease, something goes awry with the nerve ganglia at the base of the brain. This leads to inflammation of the brain as well as high blood pressure. The palsied victim has a tremor (shaking) which is most pronounced in one of the upper limbs. With the tremor comes rigidity or stiffening of the muscles. This creates a restriction of movement. The patient must reduce his activities which have a physical and psychological damage. He wants to do something, but just cannot muster enough mental and physical equilibrium to carry out the command of his brain.

Severe palsy patients have a slumped over posture and become stooped and bowed. Speech fluency is often impaired. Heretofore, the palsy patient must be confined to a wheel chair or bed. Some drugs are given to reduce muscle spasms; brain surgery is also performed in selected cases for those who could not have been helped in any other way.

Approximately one million Americans have this tragic Parkinson's disease.

Small wonder that much excitement was created when three doctors from the University of Montreal, reporting to the American Academy of Neurology cited evidence that the brains of persons afflicted with Parkinson's disease showed abnormally low concentrations of two related chemicals called *dopamine* and *sorotonia*. They speculated that a deficiency of an enzyme called *dopa-decarboxylase* might be at least partly responsible for these deficiencies.

These Montreal physicians then treated patients by giving them extracted enzymes. The results were impressive and while only temporary, they were encouraged enough to continue on to seek a permanent cure for the dreaded and presently incurable Parkinson's disease—a cure by enzymes.

Role of enzymes in mental diseases.

Future generations may be freed from mental disease if enzymes and the role enzymes play in mental and physical health are fully understood. According to a report, schizophrenic patients secrete 58% less of the enzyme cholinesterase than do normal persons.

Schizophrenia is the most common of serious mental illnesses. Often called a "split personality," it is an ailment for which hospitalization is most often required. Generally speaking, schizophrenia means departure from the real world into that of an illusion of one's own creation. Some schizophrenics look sad, others happy. All daydream their lives away; often, they hear inner voices.

Other schizophrenics act silly and talk childishly; they do not know who or where they are. Some fall into a state of stupor or muscular rigidity. There are many more who show the other extreme: they are overly excited, talkative and even aggressive. Schizophrenics do not lose their minds; their intelligence does not deteriorate—they cannot use their minds properly. Their intelligence is severed from other phases of their personality.

The latest symptom is that of a deficiency of the enzyme cholinesterase. It is believed and hoped that by simple injection or administration of this enzyme, the schizophrenic patient may be cured! This indicates how mental and physical balance is ruled by the power of enzymes.

Protection against invasion of our cellular structure.

Today, we live in overcrowded cities, exposed from morning to night to agents which affect the cellular structure—these toxic agents are introduced into your body by breathing, ingestion, injection, irradiation and even by absorption through your skin pores. You just cannot escape them.

What to do? Here is a 3-step plan for protecting yourself:

1. *Increase your catalase intake.* This enzyme, as well as many others, is destroyed by heat. This enzyme is found in raw foods. Cooking and processing will destroy catalase so it means that when you cook foods, catalase is eliminated. It would be to everyone's advantage if the consumption of fresh fruit and vegetables were to be markedly increased, thus ensuring a far greater intake of catalase and peroxidase. *Garlic* is a food that is very rich in the catalytic systems containing catalase and peroxidase. Garlic-eating people appear to be free of tumorous diseases. So—eat lots of *raw* fruit and vegetables—and garlic! Use diced and chopped garlic as a flavoring for vegetable salads. But remember—eat fruit, vegetables and garlic in a *raw* and uncooked condition.

2. *Cause your own cells to increase catalase manufacture.* The enzyme, *catalase*, is sparked to increase by intake of raw fruits and vegetables—but body movement seems to favor manufacture of this vital substance.

A normally active creature was imprisoned in a cage, with limited normal muscular activity. After a few weeks, the catalase content of the body decreased. It is very probable, therefore, that man's chronic habit of limiting his muscular activity diminishes his normal catalase level. Many other physical habits adopted by modern civilization decrease the ability of our cells to respond to these normal body stimulations which reflexively induce the production of catalase.

In brief—the sedentary person is one with a low catalase rate; this *may* account for the larger incidence of tumors among adults than youngsters. In middle-age there is a tendency to take things easy, to exert one's self less, to go about more slowly. Of course, you do not go out on a binge of athletics and physical energy. But you should give yourself the exercise you need. *Walk,* instead of ride. Wherever possible, walk to and from your job, your home, your club, etc. A nightly 30 minute walk is another way of increasing your oxygen consumption and thereby providing an environment which is favorable for catalase. Your cells and tissues need oxygen to help form and manufacture this important enzyme. (A later chapter gives you the secrets of oxygen and how to have a "fresh air cocktail.")

3. *Curtail intake of substances which inhibit or destroy the action of the catalase enzyme.* You need to recognize the pollutants in your environment and set about removing them or controlling them.

It is estimated that there are now more than 3,000 to 6,000 of all types in our food and drink available today. Many of these interfere with the catalase-peroxide balance, e.g., sulphur dioxide, sodium nitrate, sodium fluoride, certain hormones, insecticides, fungicides and dyes. . . . It is not only the *catalase* enzyme which is destroyed, but also the nutrients of the food. This enzyme is important in prevention of possibly tumorous diseases. Most of the additives are not essential, and may be harmful. Natural, organic and non-chemically treated foodstuffs will give you the substances needed to build an over-all strong enzyme system to fight for your health.

How to buy enzyme-rich foods.

When selecting foods, seek out those stores where organic edibles are sold; those which have not been covered or impregnated with toxic chemicals as a result of spraying; and those which have not been adulterated or otherwise interfered with. Most cities of average size have health food stores in which may be found an abundant supply of different fruits and vegetables that are basically free from toxic insecticides, etc. These stores specialize in stocking foods that were raised in soils which received no chemical fertilizers or sprays. Such foods are not subjected to any harsh artificial substances. Raised in a natural environment, the enzyme content in these foods is potent and beneficial.

A basic plan for maintaining your enzyme efficency.

There is no limit to the hopes enzymes present for fighting disease and maintaining good health. Their most apparent function in humans is in digestion of food. The 700 enzymes work to convert this food into forms that can be used by your body's cells. These same cells need peptides and not steak; they need lipids and not olive oil. Such changes are made by these enzymes.

To get enzymes, remember they are present *only* in raw foods. *Raw* plant materials have a great treasure of natural enzymes which are built into your own body structures. Here is a little plan to follow, based on the discoveries of the aforementioned doctors and other researchers:

1. Eliminate as much processed foods as possible. This means canned foods, commercially prepared food "mixes," cereals, baked goods, white or bleached whole wheat flour, condensed milks, etc. Substitute fresh foods—here again, health stores have stone ground whole grain bread products and naturally prepared cakes for your sweet tooth. Soya milk, made from the soy bean, is another delicious dairy substitute.

2. Your foods should be fresh. Fruits and vegetables should be in season and as newly-picked as available. Use them as soon as you purchase them. Enzymes evaporate in air and water so purchase, prepare and serve the same day. They should be *raw* wherever possible.

3. Eat fresh-killed meats *only*. Avoid any meats that have been canned, spiced, pickled, preserved, processed, dried, rolled, etc. These are all chemical-drenched and artificially flavored. They are not favorable for enzymes, neither do they contain any enzymes. Processing has destroyed enzymes.

4. Each and every meal *must* have a *raw* and *natural* food. For example, your breakfast should have a raw fruit—either a banana, apple, berry, prune, fig, etc. The varieties and combinations are endless. Lunch should have raw celery, carrots, lettuce, etc., or any other raw edible. Your dinner should have a raw salad, as well as raw seeds and nuts—sun-

flower seeds, pumpkin seeds, pistachio nuts, pecans, peanuts, walnuts, chestnuts, cashew nuts, etc. Wheat germ (raw, not toasted) and desiccated liver are other natural foods that are prime sources of valuable enzymes. These items are also on sale at specialty food shops and health stores throughout the country. Many will sell them through mail order if you write and ask them for a catalogue, price list and ordering blank.

5. When you cook vegetables, use as little water as possible to conserve enzyme and vitamin content. Cook just until tender in a tightly covered utensil. Fresh foods should *not* be soaked because water is destructive of precious enzymes. *Never* use baking soda in cooking water because it is an alkali and is destructive of vitamins, minerals and completely destroys enzymes.

BRIEF SUMMARY OF CHAPTER 3.

1. Properly used and properly utilized, enzymes are able to dissolve blood clots, bring down inflammation and swelling, facilitate cataract surgery, "erase" scars, alleviate slipped back disk, ease shaking palsy, inhibit tumors.

2. Enzymes are interfered with when subjected to harsh chemical additives, artificially administered hormones, insecticides, dyes, etc. Foods eaten should be of a natural organic source and spray-free.

3. Each and every meal should have a raw food—such as a raw salad, a raw fruit, etc.

4

HOW TO MAKE ENZYMES
WORK FOR YOU

For your mind and body to enjoy tiptop health, there must be an uninterrupted production of hormones, vitamins, minerals, amino acids, enzymes, digestive juices, tears, skin oils, and so forth. Internal and external secretions must be continuously manufactured in body processes. Yet there are no independent functions. Each one relies upon another in a harmonious family. Enzymes, powerful as they may be, require the presence of vitamins, minerals and proteins to spark their function.

Looking at your body, we know that it is made up of billions of tiny cells. Just a tiny little skin scraping examined under the microscope shows so many elements, you become amazed at the intricacy of your body. Of course, you would not be able to see the more than 100,000 enzymes distributed throughout that tiny bit of skin scraping. Tiny and almost invisible they may be—but scientists know that without enzymes, life would cease.

Enzymes are needed to control all mental and physical functions. They determine your eye color. They put strength in your muscles. They give youthful elasticity to your skin. They also determine the characteristics of your unborn children. Small wonder that the *Columbus Health Bulletin*, in describing these miracle workers, declared about enzymes that ". . . they control man as they control all life, from the instant of conception to the moment he dies, his successes and failures, his characteristics and personality, his complete destiny."

Because enzymes are responsible for digestion of food, we can appreciate their dominant values. For example, enzymes create the internal chemical action to transform a steak into tissue energy by extracting amino acids to be built into your bones, skin, nerves, muscles. Without sufficient enzymes, the steak would remain indigested to a certain extent, giving you a bad case of stomach upset and cramps—and also leading to a deficiency of proteins; and amino acids are made from proteins.

What coenzymes are.

As powerful as enzymes are, they have one weak spot—they cannot function without the presence of other substances which are known as *coenzymes*. This weakness puts enzymes under your control. It means that you are often able to stimulate enzymatic functions by making available the required coenzymes. What are these? Namely, vitamins and minerals as well as proteins which are found in both foods and also commercially prepared food supplements.

This is important to older folks since it has been seen that the ability to manufacture enzymes decreases with approaching age; it means that those who suffer from indigestion and nutrient deficiency should eat as many raw fruits and vegetables as possible since this will supply the vitamin coenzymes needed to "feed" the enzymes and enable them to function in digesting foods.

Good digestion and healthy enzymatic function actually starts in your mind. That's right—a calm and relaxed state of mind *before* you sit down at your dinner table is about the most valuable coenzyme available. Some words of advice on proper digestive health is offered by Fritz Kahn, M.D., author of *Man in Structure and Function*:

"Ideas have a greater influence on the character of the gastric juice than the food itself. A disturbing thought suffices to interrupt the digestive machinery within us. For the stomach to receive food freely, to cover it with gastric juice by means of vigorous movements, a meal must be eaten without any disturbance and with good spirits.

"The table must be appropriately set, the food appetizingly prepared and served, and the conversation interesting but not too serious. All unpleasant news should be avoided during a meal. Don't read a newspaper while having your soup, and don't think of matters to which you must attend in the afternoon while chewing your meat. The hour of eating should be an hour of forgetting, devoted entirely to the enjoyment of food. This is the best medicine for the maintenance of youth and health."

Let's just examine the enzyme and see how essential it is to treat them gently so they perform properly. Enzymes are groups of substances which serve as catalysts to perform many functions. This means that they change foods ingested, but do not undergo any change in themselves. Each of the trillions of cells in your body contains thousands of these minute substances. It is estimated that a billion enzyme molecules are involved in just one moment of activity in just a single cell. Every single body activity—mental and physical—needs its own specialized enzyme. You cannot substitute one enzyme for another. During the digestive process, each enzyme attacks a particular food and will have no influence on other foods. For example, ptyalin, the enzyme in your mouth saliva, serves to split and break down starch molecules to simple sugars. Your body cannot utilize starch in its original form. When you eat bread foods, cakes, macaroni, whole wheat products, the starch contained must be converted into these simple sugars by ptyalin, in your mouth saliva, then swallowed as such to be used by your body to create energy.

No other enzyme in your system can do the work for ptyalin. This means that if you complain of feeling a "lump in your stomach" after eating macaroni or any starch food, it is possible that a deficiency of ptyalin, the enzyme converter, is responsible for undigested foods reaching your stomach.

Pepsin, an enzyme present in gastric juices, acts solely on proteins which are broken down into amino acids. Your body cannot use protein "in the raw." When you eat meats, chicken, fish, cheeses, beans, nuts, etc., the protein content is attacked by pepsin, converted into amino acids which then serve to nourish your bloodstream, build strong skin and tissue cells and also strengthen your skeletal structure. Again, no other enzyme will do pepsin's work. If you say that steak or chicken "does not agree" with you, it undoubtedly means that you are deficient in the enzyme, pepsin.

Hydrochloric acid, another vital internal juice, is needed to break down tough meat cuts and fibrous foods to release their nutrients.

Rennin curdles milk for digestion; as a coagulating enzyme, rennin makes this curdling process so that milk and other dairy products are easily digested.

Milk, incidentally, contains both protein and fat (cream) which means that it does not combine successfully with other foods. As soon as milk enters your stomach, it coagulates—it forms curds via the enzymatic process. These curds tend to form around the particles of other food in your stomach and insulates them against other enzymes. This prevents the digestion of other foods until the milk curd is first digested by rennin. For improved enzyme health, take milk alone.

Lipase, another enzyme, breaks down fat that is already emulsified; i.e., cream and egg yolk, then changes the substance to fatty acids and glycerol.

Amylases acts upon carbohydrates. Proteases upon other types of protein, lactase upon lactose, sucrase upon sucrose and so forth.

Your pancreas also issues juices or enzymes which influence further assimilation of carbohydrates, proteins and fats. Pancreatic enzymes include trypsin—a powerful catalyst which is a protein-splitting enzyme and continues the digestion of protein. Amylopsin advances the digestion of those starches which may have escaped action by ptyalin in your mouth saliva. Lipase converts some fats into fuel to be used as energy or for storage purposes.

Enzymes must constantly be replaced.

Once enzymes complete the task they were created to perform, they are destroyed. This means that you must have a constant replacement of enzyme supply for life to continue.

Since the possibility of a gradual lessening of secretion of enzymes may be responsible for body aging and ailment, it is essential that you do all you can to maintain a normal and top peak flow of these gastric juices. And you need *all* of them.

Sure, there are more than 700 different named and known enzymes in your body. So you may let down your guard and decide that one little "lost" enzyme couldn't make that much difference. Oh no? Let's look at these two examples:

Dr. Ines Mandel of Columbia University reported to the American Chemical Society that he, together with his assistant, Betty B. Cohen, discovered that *elastase,* an enzyme, working together with a bacterial substance named *elastin,* is believed to be responsible for the elasticity of the walls of the arteries. This enzyme, still under study, may be the key to unlock the mystery of hardening of the arteries or arteriosclerosis.

Three physicians from the University of Montreal, reporting to the American Academy of Neurology, also called attention to the relationship between enzymes and another disease—Parkinson's disease or shaking palsy. These physicians cited evidence that the brains of persons afflicted with Parkinson's disease showed abnormally low concentrations of two related chemicals called dopamine and sorotonia. They speculated that a deficiency of an enzyme called dopa-decarboxylase might be at least partly responsible for these deficiences.

These physicians received impressive results by treating patients with Parkinson's disease by giving them preparations of a compound from which the body normally makes the enzyme of dopamine. Results were encouraging from the start. Perhaps the more than one million Americans suffering from shaking palsy will soon learn that enzymes are more valuable than they imagine, and that just a defiiciency of *one*, tiny, innocent little enzyme may be responsible for a lifetime of crippling disease.

Getting enzymes to work for you.

How can you make enzymes work for you? The first step is to have an adequate amount of coenzymes—the substances that stimulate enzymatic function. Here is a brief check-list of the known vitamins, minerals and proteins and their purposes in building body health and food sources. Before you go down the list, remember what we told you earlier—enzymes are found only in raw foods or those that are cooked at a temperature lower than 122° F. To obtain enzymes, your diet *must* have as many raw foods as is possible. No one is asking you to eat a raw lamb chop! Neither are you asked to eat a raw fish! But you are able to eat raw apples, raw celery, raw carrots, raw pears. All raw foods are rich in these vital enzymes. As soon as you cook any food, enzymes perish when the temperature goes over 122° F. Of course, you cannot live exclusively on raw foods so you owe it to your enzyme health to eat as much raw food as is possible. Now, let's turn to the coenzymes:

The role of vitamins.

"The vitamins are significantly concerned in *all* metabolic processes," states Herbert Pollack, M.D. in *Nutritional Observatory.* "The proper utilization of other nutrients requires a sufficient amount of vitamins in the tissues. The vitamins are involved in the catalytic phenomena of intermediary metabolism."

Vitamins are organic food substances needed for growth and reproduction; formation of antibodies; coagulation of the blood; resistance to infection; formation of intercellular substances and integrity of bones, teeth, skin, blood and nervous tissue. Vitamins further serve as coenzymes for innumerable chemical reactions concerned with the metabolism of food on which body nutritive health depends. Vitamins promote digestive processes, lactation, nervous control, metabolism. Vitamins enable your body to utilize and assimilate other nutrients and they also stimulate enzymatic health. The known vitamins are:

Vitamin A. Needed for good visual health, skin nourishment, the digestive glands to issue enzymes, the respiratory system and also the genitourinary tracts are affected by improper vitamin A. Good food sources are yellow and green, leafy vegetables, yellow fruits, kale, spinach, collard greens, carrots, pumpkin, yellow sweet potatoes, apricots, peaches, cantaloup. Found also in whole milk, butter, eggs, liver, kidney and some fish.

Vitamin B-complex. This vitamin consists of about 11 different members. This particular vitamin group is especially rich in the coenzymes —the catalysts which stimulate the bio-chemical processes to turn nutrients into energy and provide body health. The B-complex group is needed to give you a feeling of vitality, mental energy, good heart and nerve function, appetite, intestinal motility. B-complex is also needed for the oxidative processes for protein, fats and starches. Deficiency may be seen in cracked mouth corners, swollen eyelids, nervous disorder, sore and red tongue, poor skin health. Riboflavin or B-2 is needed to help the enzymes break down sugars and starches to give you body energy. Riboflavin also joins with protein to form other enzymes. Niacin, another member of this family, is needed to stimulate function of hydrochloric acid and gastric juices. Pyridoxine or B-6 is valuable for the enzyme that metabolizes protein and fat and also in breaking protein down into usable amino acids. Pantothenic acid (taken from the Greek, meaning universal) is a wholesome vitamin because it works with the protein constituent of the enzymatic chain and aids in bringing about proper digestion of all foods. B-12, a recent discovery, contains *cobalamin*, a substance needed to provide nutrition for the molecules in the bloodstream. Good food sources for B-complex include whole grain products, nuts, seeds, soybeans, corn, rice, some oatmeal, eggplant. Brewer's yeast is a prime source of the B-complex vitamins.

Vitamin C. This vitamin influences the adrenal and other glands to issue hormones which also act as coenzymes. Hormones, too, influence secretions of enzymes and need vitamin C for stimulation. Enzymes help break down vitamin C for stimulation. Enzymes help break down vitamin C to become collagen, a glue-like substance to join together cells in the bones, tissues, teeth and blood cells. A deficiency of collagen, the body's cement material, may lead to capillary damage; bleeding from spongy and damaged gums; poor teeth and fragile bone structure. Leg difficulties include painful walking and cramped sitting. Fatigue is another common symptom. Your enzymes want to give you the vitally needed collagen but cannot do it unless you give your enzymes enough Vitamin C. Good food sources are all citrus fruits, orange and grapefruit, strawberries, lemons, limes, tangerines. Vegetable sources include tomato, raw cabbage, green peppers, broccoli, turnip greens.

Vitamin D. Your bones need calcium. How can they get it? The body must take calcium and phosphorus from your intestinal tract carrying these minerals via the bloodstream to be deposited in your bones and teeth and elsewhere. But Vitamin D is needed to promote this absorption. And Vitamin D is influenced into action by enzymes in your system. It is possible that arthritis, rheumatism and other bone disorders are traced to an insufficient amount of Vitamin D and a deficiency of enzymes. Both work together. To get enough Vitamin D for your enzymes, you should look into such souces as cod liver oil, halibut liver oil, viosterol. percomorph oil (one of the richest sources), enriched milk.

Vitamin E. Enzymes cause this vitamin to unite with oxygen both within and outside your body to create a barrier against fat rancidity. If you are deficient in either Vitamin E or enzymes needed to stimulate this vitamin, it is possible that fat becomes rancid and destroys Vitamin A in your system. Furthermore, Vitamin E has been seen to stimulate heart function and build resistance to diseases in cellular activities. You find Vitamin E in wheat germ oil, whole wheat and whole grain (unbleached) products, Brewer's yeast.

Vitamin K. Enzymes will take this vitamin from foods ingested and see to it that it becomes absorbed by your body to manufacture prothrombin which forms fibrin, the main component of the healing of wounds. If you are deficient in enzymes, this vitamin cannot be properly utilized and there may be problems of fragile skin, intestinal ailments and impaired fat absorption from your intestines. Vitamin K is found in green leafy vegetables, tomatoes, egg yolk, liver, soybean oil, cauliflower.

Role of minerals.

Now let's look at minerals. Without minerals, vitamins could not do their jobs. Without minerals, enzymes lie still and the body may perish.

In speaking of minerals, Dr. Henry C. Sherman, author of *Nutritional Improvement of Life*, declares, "The body's framework or skeletal system of bones and teeth owes its strength and normal form to the fact of its being well mineralized. Smaller amounts of much soluble mineral salts are constantly present in the soft tissues and fluids of the body." As for enzymes, when they cooperate with minerals, they have the power to "put life into" the proteins in all body tissues and fluids.

Here's how enzymes rule over minerals—but enzymes need minerals to be able to do their own jobs. When minerals are extracted from eaten foods by enzymes, these enzymes create an alkaline "ash" which enters into the composition of each body tissue and in your blood. Enzymes transform these minerals into detoxifying agents to combine with acid cellular wastes, neutralizing these wastes and then preparing them for elimination. You need minerals to prevent decomposition and internal rotting, so to speak. Enzymes stimulate minerals to do something else.

The secret of osmotic equilibrium.

Enzymes create the condition of *osmotic equilibrium*. Here's how it works: your blood and lymph systems are liquids which keep solids in solution. All of your body cells are never-endingly being washed in lymph fluids. All body cells are semi-fluid with dissolved matter. Suppose the lymph outside of your cells has as much dissolved solids as the lymph within your cells. The risk is that body cells may shrink and dissolve.

But suppose you have more dissolved solids inside your cells than outside—your cells may swell up with these solids and burst! How can you keep a balance?

The answer to the secret lies in enzymes which extract minerals from foods and use these minerals to create an even balance of dissolved solids both inside and outside the cells—equalizing both internal and external pressures. This means that billions upon billions of bodycells are normal. Called *osmotic equilibrium*, this delicate balance is ruled by enzymes—but enzymes need minerals as working tools to create such a balance. Let's see some of the more valuable minerals and their functions.

Calcium. Only 1% circulates throughout all soft tissues and body fluids. The rest of the calcium is stored in your bones and teeth. You need calcium to activate certain enzymes to spirit the process of healing, blood clotting and to control fluid passage through cellular walls. Enzymes see to it that calcium is stored in the ends of bones in long, needle-like crystals called trabeculae to be used in times of stress. A deficiency of this storage may cause your body to seize calcium from your bones to be used elsewhere. If insufficient enzymes are the problem, then calcium deficiency may lead to round shoulders, brittle bones. You see, even if you have enough calcium, you need enough enzymes to generate utilization! Raw food sources of both calcium and enzymes include all dairy and milk products, cheeses, green vegetables, broccoli, kale, collards, string beans.

Phosphorus. Enzymes take phosphorus from foods and build this mineral into all body cells; enzymes use phosphorus to convert oxidative energy into cellular makeup. Enzymes use phosphorus to influence protein, carbohydrate and fat synthesis to create a normal muscular contraction, glandular secretion, kidney function and nerve impulse. This mineral has another value, according to George K. Davis, M.D., in *Nutritional Observatory.* "It is concerned with acid-base regulation and with vitamin and enzyme activity. The 20% of body phosphorus located in tissues other than bone and teeth is distributed in every cell of the body and vitally concerned with the functions of these cells. Phosphorus is tied to most of the vitamins in the enzyme systems of the body and is closely allied with carbohydrate function." A deficiency of phosphorus may be seen in weight loss, poor muscular coordination, mental sluggishness, fatigue. Phosphorus is found in dairy products, milk, cheeses, fish and poultry. Don't overlook cereals, green peas, nuts.

Iron. Here is one of the most valuable minerals needed by enzymes to bring about oxidative reactions. For example, enzymes need iron to carry oxygen all over your body to be deposited in your billions of skin tissues —including those in your brain! Since you have about 5 million red cells in one cubic millimeter of blood and iron is needed for every single cell, you cannot afford to be deficient. Enzymes will help form chromatin for cellular health—but enzymes need iron as working material. Both enzymes and iron need and use each other. Iron is found in egg yolk, organ meats such as liver, kidney, or heart. Remember the whole grain bread foods, plums, cherries, raisins (excellent iron food) and grapes.

Iodine. Your thyroid gland secretes thyroxine—a glandular extract which regulates body metabolism, growth, energy, expenditure. Before your thyroid gland can secrete this hormone, it must wait for enzymes to spark this process. And enzymes won't work by themselves—enzymes need iodine to perform glandular action. Enzymes have already extracted myoglobin from iron-rich foods; now enzymes will take myoglobin and blend it with iodine to serve as a storage site for oxygen in your muscle tissues. But again—iodine, in itself, cannot perform this function without enzymes. And enzymes cannot do the work without iodine—extracted from such foods as iodized salts, vegetables grown in iodine-rich soil, onions, red cabbage, asparagus, mushrooms.

Sodium. Enzymes take sodium from your foods and help keep the delicate water balance between cells and fluids. Enzymes blend together sodium and chlorine, another mineral, to improve your blood health. Sodium is found in meats, fish, sea foods, beets, carrots, dandelion greens.

Potassium. Works together with sodium to normalize your heart beat and give nourishment to your muscular system. When enzymes have both sodium and potassium, they are able to transport oxygen to your brain cells. Enzymes see to it that sodium is left in the fluids circulating *outside* your tissues and a tiny amount inside. Enzymes also see that potassium is found mostly *inside* your tissues and just a small amount outside. Enzymes maintain this very delicate balance. Potassium is found in all citrus fruits, watercress, mint leaves, asparagus, chicory, green peppers.

Magnesium. Magnesium plays an important role as a coenzyme in the building of protein, according to Dr. Ruth M. Leverton. "There is some relation between magnesium and the hormone cortisone as they affect the amount of phosphate in your blood. Animals on a diet deficient in magnesium become extremely nervous and give an exaggerated response to even small noises or disturbances. Such unnatural sensitiveness disappears when they are given enough magnesium." This mineral is found in legumes, nuts, cereal grains, milk, cheeses, cocoanut.

You need all minerals. Enzymatic function becomes impaired or reduced if there is a deficiency in just one mineral—just as the body experiences disruption if you are deficient in just one enzyme.

Rose's *Fundamentals of Nutrition* states, "The chemical elements which make up the body sustenance must be nicely balanced or trouble ensues. The efficiency of each mineral element is enhanced by proper amounts of the others."

Enzymes as proteins.

Enzymes regard proteins as the "main dish" they need for function. You are nearly all protein. Your muscles, skin, hair, nails, eyes are protein tissues. Blood, lymph, heart, lungs, tendons, ligaments, brain and nerves are also protein. Hormones, the enzymatic stimulated regulators of body processes are proteins.

Enzymes, themselves are proteins!

When enzymes get hold of proteins which they then transform into amino acids, they are able to do a major job of keeping you healthy and happy. Remember: your body *cannot* use proteins in the raw, so to speak. Your body must convert protein into amino acids—by enzymes—and these substances are then used to heal wounds, build resistance to infection, strengthen your liver to filter out toxic agents, enable you to recover from illness. Enzymes take certain amino acids to build a healthy bloodstream. Enzymes bring into play all of your body hormones to blend amino acids together. Enzymes send one amino acid to your bloodstream to make red blood cells. Enzymes send another amino acid to give you strong fingernails. Enzymes convert protein into hemoglobin which enters your bloodstream and facilitates proper oxygen transportion from head to tip toe. Still, enzymes will use another amino acid to regenerate damaged brain tissues. Hormones need amino acids so enzymes extract one or two of these and send them to your glands.

But again—enzymes are victims of one weakness: they themselves, need amino acids so will need some protein for their own strength. And since enzymes are also protein substances, you can readily appreciate the values of eating good foods to give them their own nourishments. Protein rich foods include milk, beef, calves' liver, fish, eggs, Brewer's yeast, nuts, seeds, grain, soybeans, rice, oats, wheat and millet. Egg yolk is valuable protein: it is rich in iron compounds and fatty proteins needed to build strong nerve and bloodstream health. Egg white and egg yolk are high on the protein source list—containing 97% of all amino acids— there are about 25 known amino acids and countless unknown ones.

What enzyme environment means.

So, the most vital step in creating an *enzyme environment* is to supply them with vitamins, minerals, proteins and other essential nutrients. Food supplements are valuable in situations where you may be eating a balanced diet but cooking, exposure to air, water, storage, freezing etc., deplete the delicate vitamin-mineral content. To guard against nutritional deficiency, an all purpose supplement is helpful and a "dividend" to your enzymes. After all, enzymes have to eat, too.

How to keep your enzymes healthy.

In these chapters, you were told how certain food combinations favor and destroy enzymatic function. You are able to improve enzyme health by following some simple health rules. Here are 13 golden rules that will give health to your enzymes and enable them to properly function in building health for you:

1. The main, hearty meal should be eaten earlier in the day, preferably at noon. Avoid eating heavily before going to sleep. Enzymes, when working to digest your food, call upon a flow of blood to your digestive organs. There is an inflation of the blood vessels in the digestive tract—and this causes a subsequent, but natural, constriction of blood vessels in other body parts since the greatest need for blood is in this digestive system. All parts of the body cannot be supplied with blood at the same time. If one part gets an extra blood supply, some other part must get less. During sleep, blood is withdrawn from the brain and muscles to be sent to internal organs that continue functioning while the body is at rest. Since the brain is already deprived of some blood needed for digestion, further depletion sent away from the brain during sleep, may create a health risk. So, eat a main meal earlier in the day to give a natural blood balance in the body. Since blood is a form of protein needed by the enzymes, take care not to deprive one part of your body of this valuable substance. Sleep, incidentally, is sounder if digestion is performed long before you retire.

2. Eat slowly and chew thoroughly. Carbohydrates and starch foods digest better in an alkaline base. Your mouth saliva is an alkali base; the enzymes secreted by your salivary glands are alkaline and are issued forth only when you chew thoroughly. To bolt down foods means that the ptyalin, the salivary enzyme, has not been given a chance to fully digest the starch which reaches your upper intestinal tract inadequately prepared for overall assimilation. Chewing also has a powerful influence on your digestive tract, causing enzymes to alert and make ready to receive swallowed foods. Chewing, too, enables food to reach the digestive tract in a finely pulverized form, easily handled by enzymes, preventing gassy indigestion.

3. Restrict your tea, coffee and cola drinks because of caffeine. These drinks, taken to excess, bring a lot of caffeine into your digestive tract; caffeine may irritate your stomach and redden the gastric mucosa. Often,

caffeine causes an irregular heart action and has an interfering reaction on enzymes. Caffeine is hostile to some enzymes.

4. Avoid spices and condiments such as pepper, mustard, vinegar, catsup and pickles. They irritate the enzyme-forming ducts along the alimentary tract and also disturb the liver and kidneys. Some enzymes become weakened by harsh spices.

5. Substitute freshly squeezed lemon juice for harsh white vinegar when making salad dressing. Fresh, pure, unadulterated apple cider vinegar is a boon for stimulating enzymes. Mix equal parts of this apple cider vinegar with vegetable oil for a heavenly salad dressing. Harsh white vinegar contains acetic acid; in concentrated form, it is powerful enough to remove warts and moles. It may also destroy enzymes. It's natural for you to have a craving for sour flavors. Satisfy it with citrus fruits and lemons used in moderation.

6. Eat vegetables raw, wherever possible. Enzymes in vegetables are destroyed by high cooking temperatures. Steaming is also destructive but is the lesser of the evils. Eat as many vegetables as possible—in the raw.

7. Avoid eating highly acid fruit at the same meal with coarse vegetables. The combination may lead to gassy indigestion. Fruits require little enzymatic action for digestion; when fruits are kept in the stomach until the vegetables (which need enzymes) are ready to pass through, this improper combination may lead to fermentation. This is frequently regarded as indigestion. Eat fruits an hour or two before you will be eating vegetables.

8. Reduce intake of custards or rich puddings which contain combinations of milk, eggs, cream and sugar. Together, this is a rich mixture which may disturb digestion.

9. Foods should be baked or boiled rather than fried, for better enzymatic action. Frying coats food with grease and prevents enzymes from attacking it, rendering digestion difficult or impossible.

10. Avoid drinking heavily *with* meals. Large amounts of liquids of any sort will dilute the enzymes (which are liquids, too) and weaken their powers. Drinking with meals leads to the bolting down of food habit. Instead of thoroughly chewing and insalivating your food, you tend to wash it down half-chewed. Since enzymes work best with thoroughly chewed foods, avoid the habit of drinking with meals. A little sip of room temperature liquid is permissible for very dry foods. Cold drinks—

whether water, lemonade, punch, iced teas, etc., are often consumed with meals. Cold stops enzyme action! The enzymes have to wait until the stomach temperature is raised again to normal before they can resume their action. A cold drink going down the stomach has a shocking and chilling effect. It causes a peculiar feverish condition leading to great thirst. Hot drinks weaken and enervate the enzyme producing glands in the digestive tracts. Hot drinks destroy stomach tissues and weaken power of enzymes. The functional powers of the enzyme forming glands are best when working in a temperature that is similar to normal body temperature—or, not more than 100° F. Here's a good rule of thumb—drink your liquids fifteen minutes before meals, thirty minutes after fruit meals, two hours after starch meals and four hours after protein meals. Long after you have swallowed your food, your enzymes are still at work digesting it—so give them time!

11. For cheeses, select cottage cheese, Philadelphia cream cheese or similar. Shun processed cheeses. When commercially processed cheeses are ripened, there is a development of offensive bacteria. In cottage or Philadelphia cheese types, the ripening is natural. These cheeses contain lactic acid bacteria which tends to prevent putrefaction and also inhibits the formation of harmful by-products which inhibit enzymatic function.

12. Avoid processed and refined foods such as sugar, white bread and all white-flour products such as polished rice, white macaroni, thickly peeled potatoes. Processing has depleted valuable vitamin and mineral content. Enzymes need these nutrients to work with and for their own powers. Select foods that are as natural as possible.

13. Simplify your meals; that is, avoid complicated mixtures or varieties at one meal. Avoid eating between meals. Unlike the stomach of a cow or horse (animals that munch all day), your stomach should utilize one meal, then have a rest period before taking on another batch of food. If your stomach is not given this respect, it creates an effect such as the milk crock not cleaned between milkings. Your stomach goes sour and you go sour with it. Fruit juice, requiring small enzymatic secretions, is an exception as it leaves the stomach speedily.

How to complement your enzymes.

You will want to complement your enzymes. It is true that foods contain enzymes but you want a special supplement to make up for any possible deficiencies. This means that you need those foods which are particularly plentiful with enzymes—called exogenous enzymes. (Those produced by the body are called endogenous enzymes.) The more exogenous enzymes you get, the less body enzymes you need and the more favorable digestion becomes.

There *are* special foods which you can obtain to give you valuable enzymes. You can mix them in a delicious and health-building drink, right in your own home. This Special Enzyme Cocktail, when taken 30 minutes before mealtime, will supercharge your entire digestive tract and give you a healthy appetite—and also provide a veritable treasure supply of enzymes. Here are the foods:

1. *Papaya.* Visit any health or special diet food store for pure, unsweetened and unadulterated papaya juice. The papaya is a delicious, tropical melon-like fruit known to be a wonderful digestive aid. Papaya is one of the few fruits containing a tremendous amount of papain, the enzyme needed to turn protein into usable amino acids. Papaya is also one of the rare fruits which exerts its action in an acid, alkaline or neutral media. It is highly lauded for being a rare alkaline fruit—rich in pectin, a substance which soothes the intestinal tract. Natives of the tropics call the papaya a "fruit of the gods" and lore tells us that it was used for generations as both a staple food or dessert and has been seen to overcome stomach and intestinal disorders. No doubt, the tremendous papain potency is the secret to the power of the papaya. Get a bottle of the juice that has been extracted from the entire fruit—tree-ripened, of course—including the skin for the best of the papain. It's a refreshing taste sensation and also a health builder.

2. *Brewer's Yeast.* Locked within a tiny yeast cell are the prime sources of the B-complex vitamins which activate enzymes. This vitamin family is the most valuable if any comparison has to be drawn. B-complex is the best food you can give to enzymes and also the best source of enzyme-producing agents. The cell of the Brewer's yeast may be the tiniest unit of living matter in the whole universe; it measures 1/4000th of an inch across. Yet it has such a tremendous power with its enzymes that it is regarded to be a "must" food for the health seeker. Dr. Clive M. McCay in *Yeasts in Feeding,* states that Brewer's yeast was able to *double* the lifetime of test animals who ate it over a prolonged period of time. Special diet shoppes and health stores sell Brewer's yeast in capsule or flake form.

3. *Tupelo honey.* Deep within the primitive, almost virgin territory of a river swamp along the Apalachicola River in northwest Florida, grow the astonishing tupelo gum trees. The blossoms of this gorgeous tree

resemble little doll's cushions of a chartreuse color. These blossoms are food for hundreds of thousands of bees who feed upon this pure blossom, undisturbed by chemical sprays or fertilizers, and turn the little rich nectar dewdrops into prized tupelo honey. Only in this exotic river swamp does the pure tupelo gum tree flourish in pure health. Honey made from such bees are rich in natural sugars. The most valuable aspect of tupelo honey is that it does not granulate! Bees making such honey are so unusual, they could be fantastic. The tupelo honey bee hibernates throughout the cold winter! Only during the months of April and May when the tupelo flow is high, are bees permitted to feast on the blossoms —within 4 to 6 weeks, hundreds of thousands of pounds of honey are so made. The tupelo honey bee works furiously—many even wear out their wings. At the season's end, apiary owners take the bees to a warmer climate for a rest. No other honey is so manufactured.

What makes tupelo honey so favorable to enzymes? All honey contains enzymes. But in processing, most manufacturers will filter honey which means that it must be heated to 135° F. Then it goes through an automatic filtering machine at 120° F. Why? Because practically *all* honeys will granulate and this processing makes them granulation-free for a certain period of time. Since enzymes are destroyed at such high temperatures, honey won't give you such valuable digestive help. This does not include tupelo honey which is *never* heated, *never* filtered because it *never* granulates. It is made by the healthiest bees in the world, from trees grown in rich soil in the virgin hardwood swamps of the one and only Tupelo County of Florida. For enzyme health—use tupelo honey.

4. *Rose hips.* These are relatively unknown. Rose hips are the mature fruit of the rose when the petals drop off. Ripe rose hip granules are bright scarlet in color. One of the most highest concentrations of Vitamin C is found in rose hips. Also present are calcium, phosphorus and iron. But—rose hips are valuable for still another reason. This unique natural food has a special harmony by sparking biocatalyst action—in other words, rose hips contain biocatalysts or enzymes which your digestive system needs. Rose hips are available in powder or granule form as well as in tea bags.

My special enzyme cocktail.

Now . . . here is your special Enzyme Cocktail which will stimulate your own sleepy digestive juices and also give added enzymes to your system.

Add four heaping tablespoonfuls of fresh Brewer's yeast to one glass of papaya juice; fold in one teaspoon of tupelo honey. Stir vigorously or use an electric blender. Before drinking, sprinkle with one teaspoonful of rose hips powder. Drink this enzyme cocktail 30 minutes before mealtime and you will discover that it has a dynamic power in aiding digestion and building "instant health."

CHAPTER 4 HIGHSPOTS YOU NEED TO REMEMBER.

1. Just a tiny skin scraping examined under the microscope may show more than 100,000 enzymes.

2. Enzymes depend upon good digestion for full efficiency, so eat in a calm environment, in a calm state of mind.

3. Each enzyme acts upon a specific food. A deficiency of one enzyme means you are denied partial benefit of the whole enzymatic system.

4. Vitamins, minerals and proteins join hands with enzymes; all work together. Each needs the help of the other.

5. Eat slowly and chew thoroughly. Restrict caffeine drinks such as coffee, tea and cola pop beverages. Avoid spices. Eat raw vegetables. Avoid drinking heavily with your meals.

6. Organically raised foods should be purchased, prepared and eaten in as natural a state as possible.

7. Powerful enzyme foods sold at health food stores include: papaya, Brewer's yeast, tupelo honey, rose hips.

8. Drink the special Enzyme Cocktail about 30 minutes before mealtime.

5

BEWARE:
ENEMIES OF ENZYMES

Enzymes destroyed by cooking and processing.

You are undoubtedly aware of the superior health enjoyed by wild animals. When captured and placed in a zoo, their health begins to decline. Shiny coats lose their lustre; bright eyes grow bleary. It is also well known that animals in captivity are usually sterile. Could enzymes and diet be responsible? Yes, when we learn that captured animals are often given cooked food with the results that they soon begin to show the same ailments as man. Today, most zoos endeavor to feed *raw foods*, as much as possible, to their animals. Consequently, the disease rate of zoo animals has dropped recently.

Enzymes are destroyed by high heat temperatures as well as those of processing during canning, bottling, freezing, preserving, etc.

Safe enzyme temperatures.

A slight increase in heat will help enzymes perform more rapidly. But, excessive heat will destroy them; even after the food cooked has been cooled down, the enzymes will remain inactive and useless.

When food is cooked to a point above 122° F., then the enzymes are permanently destroyed. Water boils at a temperature of 212° F., so this means that when you boil fruits or vegetables, all enzymes in these foods will be destroyed.

Meat, roasted at a temperature of 200° F. to 300° F. will, of course, result in complete destruction of enzymes.

Therefore, if you want to derive the most from the enzymes, it means that your foods should be as raw as possible. You cannot eat all of your foods raw, especially meat foods. But you can eat as many (if not all) of your fruits and vegetables in a raw state.

Refrigerate fruits and vegetables.

The enzymes in these foods will remain intact, and inactive, as long as you keep them in your refrigerator or freezer. During the growing and ripening process of a plant, enzymes are within, working busily to form vitamins and minerals and creating changes to make that plant both palatable and nutritious. As soon as the food is picked, there is a halting of the work of the enzymes. You can see this in action by the wilting of lettuce and decaying of leaf bearing vegetables. The skins of fruit will wither and discolor.

How can you arrest this decaying of enzymes? Immediately refrigerate all fruits and vegetables (except bananas) to keep the enzymes in a state of suspended animation. Remove just before preparing and serving.

Supplements containing enzymes.

Natural food supplements should make up for a substantial portion of enzymes lost in cooked foods. True, you will eat your plant foods in a raw state, but nobody is asking you to eat raw meat or fish. Your digestive system is too adjusted to a cooked food diet in which extreme heat breaks down the cell walls of the edible. So you look to such valuable enzyme-rich supplements as Brewer's yeast, desiccated liver and wheat germ oil. In supplements that are desiccated, the temperatures which processed the food should be no higher than 122° F. Local health stores contain these supplements in the form of tablets, powders, granules, oils, even in flour to be used in baking . . . although this process will again result in enzyme depletion.

Animals thrive on raw foods.

The noted Francis F. Pottenger, Jr., M.D., writing in *American Journal of Orthodontics and Oral Surgery*, tells of tests conducted on 900 cats, divided into groups.

The first group was fed raw meat and raw (unpasteurized) milk. These cats showed signs of tiptop health, produced healthy litters of kittens from generation to generation.

The second group was given cooked foods exclusively, which included milk that was either pasteurized, evaporated, or sweetened condensed. Before long, these cats developed the so-called civilized diseases: teeth were lost, fertility was impaired, birth labor was painful, nervousness increased, allergies and infections developed; they had diarrhea, pneumonia, heart and kidney trouble as well as thyroid glandular disorders and diseases.

Films shown of the cats showed the first generation to be healthy and sleek. Even when dropped from an average height, they fell with nimble motions. The second generation of these cooked food cats toppled over, when dropped; they did manage to right themselves up. The cats also had sway backs and flat feet. (Whoever heard of flat feet in animals, but it is true!)

The third generation of cats fed on cooked food could not even get up when dropped. They had scores of ailments. They looked sick.

This film brings to mind the image of a man struggling down the street, bent over, looking exhausted and sick, as if the elements of the world were against him. A typical "third generation" specimen of cooked food.

Dr. Pottenger then directed that this last generation be given an exclusive raw food diet. Before long, rejuvenation took place, especially in those where degeneration had not progressed too far. The cats became alert; their symptoms slowly vanished. They became strong, healthy, virile and fertile.

Begins meals with raw food.

The Russian scientist, Kouchakoff, learned that when cooked food is eaten, there is an increase in the formation of white corpuscles in the intestines at the end of the meal. These corpuscles are needed to fight infection, but there are times when they multiply too rapidly, thereby creating a cancerous condition such as leukemia. Raw foods, however, were found to keep the white corpuscles at a normal and healthy count. This means that each meal should *begin* with a raw food. Salads, especially, should be the first foods eaten, prior to cooked items in the same meal.

Prepare a fresh vegetable salad of shredded carrots, cabbage, chard, broccoli, onions, celery, turnips, etc. Raw and uncooked, they should be eaten *prior* to your meal. Eat them in whatever combinations you prefer, or singly.

Dr. James B. Sumner, writing in *Secret of Life Enzymes*, feels that your "fortyish" look, sagging skin, middle age spread and attitude may be because you are suffering from an enzyme shortage. With middle age, there is a temptation to eat more and more cooked foods. Problems of chewing, poor teeth condition, tiredness all lead to the "easy" practice of eating foods that are soft and palatable. Consequently, raw foods such as fibrous fruits and vegetables are virtually absent from the diet. A shortage of live enzymes invariably contributes to premature age.

Enzyme shortage makes you old before your time.

Tests conducted on middle age folks revealed that because they ate few or little raw foods, they were severely short on enzymes. Year after year, according to Edward Howell in *Status of Food Enzymes in Digestion*, these people developed skin wrinkles, thinning hair, sagging of figure and muscles as well as eyes that were blank and without luster. In other words, *you look worn-out because the processes of your body are starved for enzymes!*

Drink your enzymes.

Yes, you can drink your enzymes. H. E. Kirschner, M.D., author of *Live Food Juices*, demonstrated the vitality-building benefits of raw vegetable juices. Countless patients were snatched from hospitals and even death by being placed on an enzyme-rich program—which called for freshly squeezed juices from raw vegetables (and fruits) to be taken as much as possible.

Since enzymes are destroyed by processes of cooking, pasteurizing, pickling, smoking, canning, it means that canned juices will not provide as many enzymes as you need. Therefore, drink your enzymes by freshly squeezing juices from fruits and vegetables—drink these juices before each meal. That's right. Before breakfast, you drink a glass of a freshly squeezed fruit juice. Before lunch, drink a glass of a freshly squeezed vegetable juice. Before dinner, have a mixed vegetable cocktail. As a nightcap, you have a mixed fruit juice drink. At the same time, remember that each meal begins with a raw fruit or raw vegetable salad. This gives you a treasury of enzymes throughout the day, from the time you awaken until the time you go to sleep.

Nature's own cooking process.

Nature has a way of "cooking" plant foods so that they are not regarded "raw" but still have a wondrous supply of enzymes. Nature "cooks" a luscious bunch of purple grapes swinging to and fro in bowers of green. There is the hickory nut that has ripened in the top of a mountain tree, whose life-giving properties have been filtered through a hundred feet of clean, white wood. There is the delicious apple, or peach, reddened, ripened, and finished—nursed in the lap of Nature, rocked in her etherial cradle and kissed from the fragrant blossoms of infancy on to maturity by the soft beams of the life-giving sun, ready to be eaten. Such foods are not "raw." They have been cooked by Nature, but with no loss of valuable enzymes. Tree-ripened and sun-ripened fruits and vegetables are "cooked" foods that you must have for enzyme powers.

Steel worker victim of chronic fatigue.

Young Barry M. is a healthy, rugged, husky steel worker in a Pittsburgh foundry. He puts in a lot of overtime, doubles up during rush schedules, performs heavy manual labor. He looks healthy and rugged. Yet there are times when he is on "sick leave." With alarming frequency, he has to quit work a few hours ahead because of recurring fatigue. When examined by the plant physician, young Barry told about his eating habits.

"My wife feeds me lots of heavy meat and potato dishes. I eat fish about once or twice a week. Up until lately, I've been feeling okay. Now I'm getting weak. I guess all I need is a thick steak and french fries and I'll be ready to go again."

The physician took blood tests, noted that the enzyme content was dipping. "What about fresh fruits and vegetables? Eat much of those?"

"Sure—I know all about bunny foods," he joshed. "My wife gives me stewed fruits and vegetables nearly every day."

Now the doctor had a clue to the mystery of Barry's recurring fatigue. "Don't you eat *raw* plant foods?"

He looked blank. "No—don't like the taste of them. What's the difference? Vegetables are vegetables, aren't they?"

"Not so," said the doctor. "If they're raw and uncooked, they have certain elements that you need to give you strength and a healthy mental alertness. Cooking destroys those elements." Then he went on to explain about the extreme delicateness of enzymes and how cooking would weaken and completely render them useless. The plant doctor asked Mrs. McMahon to come in for a talk. He prevailed upon her to serve as many fresh foods as possible—and to give raw juices with each meal. "Tell your husband to chew his foods . . . to stimulate the enzymatic flow. If he's reluctant to change, do it gradually. Give him raw celery stalks, carrots, radishes, and other tubers."

No other changes were made. Barry was still enjoying his thick steaks and luscious meat roasts; but he found increasing amounts of raw and chewable foods on the vegetable dish beside his main plate. Just three weeks of raw foods helped him overcome chronic fatigue and he reverted back to a healthy and rugged state.

Women's club lecturer grows faint.

Susan W. is well known in her locality and throughout her state as a lecturer on woman's role in world affairs. She is frequently called upon by the P.T.A. to deliver talks on improved school conditions. She feels a little vain about her appearance; she diets rigorously to keep from getting overweight. Her day begins with black coffee and toast. A few hours later, she's on the podium, delivering a lecture.

Shortly before noon, one day, she was in the midst of a friendly debate when the faces of her audience began to blur. There was a strange pounding at her temples. She felt herself grow cold and clammy. Beads of perspiration broke out on her forehead. The words she spoke (and heard) were indistinct. From a vacuum, she heard the chairlady ask, "Mrs. W., are you feeling well?" Then everything went black.

She revived on a cot in the first aid room of the school where she was lecturing. The public health nurse had taken her temperature, then her pulse and also a sample of her blood. "Have you been dieting?" asked the nurse.

"Well," Mrs. W. struggled to sit up and sip the cool fruit juice given to her, "I've always dieted. But I do eat a balanced meal." She then proceeded to tell of her sparse diet. True, it contained many raw fruits and vegetables, but she ate sparingly of heavier foods, feeling they would give her unnecessary calories.

After getting a clearer picture, the public health nurse commented, "You are denying yourself valuable proteins, vitamins and minerals—you get these from meats, fish, eggs, oils. Without these nutrients, the enzymes from raw foods are weak and powerless. It's more than likely that your condition is growing weak because of this imbalance."

When Mrs. W. learned to improve her cooked food intake—and step up her raw plant food intake to keep the balance—she found herself more alert, vigorous and no longer prone to dizzy spells. She later admitted there were many times, at home or out of doors, when she previously felt weak at the knees and unexplainably nervous.

This illustrates how enzymes, in themselves, are not fully independent. Raw plant foods are the foundations upon which health is built—but enzymes function when given a spark-plug by nutrients found in cooked foods. The situation is also reversible—vitamins, minerals, proteins, carbohydrates, fats, etc., are electrified into action by enzymes. Nature has created this balance.

You need cooked foods.

Without a doubt, there was a time in history when primitive man lived exclusively upon, raw, uncooked foods. And the primitive man was a healthy brute, free from those diseases that later caused plagues and rampant destruction. When he learned of fire-making, the savage started to cook. Before this, he was brought up with the idea that food was eaten raw and no other way. Now he began to change.

In due time, he alternated between raw and cooked foods; often, he ate cooked foods exclusively because of greater palatability and ease of chewing and swallowing. When we examine skeletal remains of cave dwellers, we discover that they did have diseases—especially arthritis. Such remains are usually found at the site of prehistoric camp fires with burnt animal bones and wood ashes. We assume that the cave man of this era subsisted largely on cooked foods. He did not know about edibles (and possibly could not find them in his part of the prehistoric world) and slowly confined himself to edibles that had to be cooked since animals and fish were usually everywhere.

Yet, diseases may be traced to an imbalance of raw and cooked foods—*and an improper way of cooking foods.* You *do* need cooked foods. Many raw foods are chock full of nutrients—and enzymes—but these are locked in the tough, fibrous cellular walls which require much time for your digestive fluids to break through. This means that you must chew and chew—and then your digestive system must attack these cellulose fibers and break them to release the nutrients. *Often this breaking up may not take place with certain raw foods.*

Cooking is accomplished by using heat to annihilate the cell walls and permit the speedy penetration of the nutrients by the digestive system. Unfortunately, enzymes will be destroyed by the heat . . . hence the importance for knowing how to cook so that the rich nutrient store will be spared to be utilized by your body.

Conservative cooking.

All plant foods are to be stewed in their own juices or served with the water in which they are cooked in the form of gravy or sauce. You may steam or bake those plant foods which must be subjected to cookery—but conserve all natural flavors by this process. Use as simple a method of cooking as possible. How foolish to cook a peach or an orange (which should not be cooked) and then try to delude yourself into believing that you have improved its delicious flavor or increased its dietetic value.

Cook what must be cooked as briefly as possible. Eat what can be eaten raw in as natural a state as possible.

Use rapid cooking methods.

When you use rapid cooking at a high temperature, you produce less damage. Low heat that is long continued will cause more food damage. For this reason, forms of slow cooking are most undesirable. Steam pressure cooking depletes both enzymes and vitamins in foods—minerals, too, are washed and evaporated away.

Selecting plant foods.

Wilted greens are poor foods. Fresh plant foods are prime sources of nutrients and enzymes. The process of wilting, extracts the best nutritive elements from these plant foods, impairing their values. Animals are smarter than humans when it comes to food selection. Place wilted lettuce together with fresh lettuce before an animal and he will select that which is fresh. He uses wilted lettuce for bedding or to trample upon it. The animal may crunch the wilted food while eagerly devouring the fresh.

Preparing plant foods.

Because enzymes are so perishable, you do not soak vegetables in water for long periods of time. This depletes nutrient store and also makes them less tasteful. As soon as vegetables are purchased, wash quickly in cold water, taking care not to bruise them. Wrap in a damp cloth to protect them and place in the refrigerator until serving . . . which should be as quickly as possible.

Plant foods should not be permitted to stand in water since this depletes them of enzymes which are extracted by the water. You eat nutritionally impoverished vegetables.

Fruits and nuts.

Fruits must be eaten raw. This also applies to dried prunes, figs, peaches, pears, apples. For an enzyme-rich fruit surprise, wash the dried fruits carefully, place in a deep bowl, cover with lukewarm water. Place a lid on top of the bowl and let stand overnight. When serving, the water in which they have been soaked should be served with them—because this water is brimming with enzymes and minerals. No sugar should be added. Fruits thus prepared are delicious and more flavorful than cooked fruits, and also more speedily digested. Cooked fruit, incidentally, calls for more digestive power than most folks have. Eat raw fruits. As for nuts, these, too, should be raw, never cooked or roasted.

Cooking vegetables.

Not all vegetables can be eaten raw. Some have to be cooked. Therefore, cook them in their own juices with just enough water to prevent burning. Serve with the liquid remaining. Leafy vegetables should be steamed or cooked just enough to make them palatable and *not* until the color is changed. *When this happens, the enzymes have been destroyed.*

Cook vegetables as short a time as possible to preserve the enzymes as far as possible. Eat immediately after cooking. Never cook ahead and let vegetables stand for hours before eating. Oxidative forces are at work and enzymes, together with vitamins and minerals, will evaporate. Twice cooked vegetables have less enzyme value and are less digestible than once cooked vegetables. Waterless cookers are highly desirable.

Carrots, beets, turnips and other tubers and fruiting plants such as squash, tomatoes, etc., should *not* be pared or cut up prior to cooking. Scrub with a brush and cook them whole. Serve and eat whole, flavored with a little butter or vegetable oil. As for potatoes, cook them in their skins; eat with the skins. Bake them 40 minutes in a very hot oven or steam them in a waterless cooker.

Animal foods.

Eggs should be poached, coddled or soft boiled. As for meats, you would be wise to bake or broil under the flame to retain their juices in which are located valuable proteins. Serve meat with its own juices as natural gravy. Fish may be steamed or baked. Never fry meats because this process breaks open internal fat globules which have an inhibitory effect upon digestion and also will hinder enzyme absorption in your body.

Cereals.

Avoid boiling cereals of any sort. Dry cereals are good when served with fresh slices of fruit and milk. Do you like toast with your morning cereal? Actually, browned bread is a mixture of charcoal, tar and ashes. When over-toasted, nothing but ashes remain. You would do better to eat stale bread, than toasted bread.

Soups and gravies.

The one major fault of soup is that you "eat" it without chewing and this does nothing to stimulate the flow of enzymes. You might put bits of meat in soup as well as slightly cooked vegetables which require chewing. Clear soups, or bland concoctions should be sipped very slowly. It has been seen that inhaling fragrant soups will spark the flow of digestive juices and will be helpful to the ensuing enzymatic process. A half cup of clear soup might thus be utilized for those who must have this aperitif. As for gravies, if they are *natural*, they are suitable. Artificial gravies are usually ladened with starch, flour, tapioca, harsh condiments which are destructive of enzymes.

Dried foods

Today, we see a wide variety of dried foods on the market, other than fruits. We see milk, vegetables, meats, etc. At one time, it was believed that the process of dehydration deprives these foods of nothing but their water. This has since been found to be untrue. The dehydration process calls for a comparatively high temperature and this destroys perishable vitamins, minerals and enzymes.

Cabbage, for instance, is an excellent source of enzymes and vitamin

C which work together. Yet, dehydrated cabbage requires a temperature that destroys both vitamin and enzyme supply. Furthermore, foods that are dried are then subjected to long storage periods. That's why they were first dried—to facilitate storage. This, too, causes evaporation and destruction of prime nutrients and enzymes.

Bleached foods are also enemies of enzymes. Baking soda or other alkali work speedily to destroy these life giving elements. Drying and processing calls for fats, grease, salts, soda, acids, sugar, spices and pungent extractives which interfere with digestive powers of enzymes and even stimulate to overeating.

Sulphured fruits appear to be saturated with poisonous sulphurous acid that is used in bleaching them. These are sold as "dried" fruits, when actually, they are chemical-dried and chemical-drenched. Harsh chemical action is highly destructive of enzymes so you are eating fruits that not only contain possible contamination, but are valueless insofar as enzymes are concerned.

You may have heard of sun-dried and dehydrated fruits that are prepared by natural means. Apricots, raisins, dates, plums, prunes, figs, etc., are a few of such edibles that have been dried in Nature's own sunshine. Sold at special diet shops in cellophane or paper bags, they are safest and most nutritious. The label usually tells you that no chemical preservatives or insecticides have been used. Such dried fruits and vegetables are rich in enzymes because they have not been subjected to harsh artificial methods. They have not undergone what is called a "denaturing" process.

Since prolonged heat is in any case injurious, it is obvious that drying of foods at an artificially raised temperature must be extremely disadvantageous from an enzyme viewpoint.

Frozen foods.

As stated earlier, freezing will inactivate enzymes, keeping them in a condition of suspended animation. This is wonderful! If you can have your own freezer, this is the answer to your problems during the cold winter months when many fresh fruits and vegetables are unavailable. You can hurry all of your luscious plant foods, during the season, right into your freezer to be kept there in cold months. But remember: enzymes will begin to disappear with thawing so eat frozen plant foods as soon as they are thawed out.

Cooking with supplements.

Certain enzyme-rich preparations for cooking are available. Take advantage of them. Nationwide, health food shoppes sell cold-pressed wheat germ oil. This means that the process by which oil was extracted from the whole wheat kernel was kept at a temperature that is not harmful to enzymes or several valuable vitamins. Use this type of wheat germ oil wherever you need lubrication—that is, to brown or braise meat prior to stew-making, to steam liver for baking, to add to baking vegetables or casseroles. The wheat germ oil provides an excellent source of valuable nutrients and enzymes.

Peanut butter is another good source of enzymes, especially if it is free from harsh additives and added ingredients and prepared at special vacuum temperatures which temporarily "knock out" enzymes. Here again, health stores carry different types of nut butters which you may find fascinating by variety. A delicious enzyme pick-me-up is a sandwich made from natural peanut butter, whole wheat bread, apricot slices and a glass of freshly squeezed lettuce and carrot juice.

Whenever a recipe calls for flour, select stone ground flour at your health food shoppe. The same applies to bread products, flavoring agents such as vegetized salt and sweet rose hips. Select these products, made from a natural base, at local shoppes.

What did you eat for lunch?

Young Timmy was embarrassed over the outbreak of pimples on his face. At 16, he had arrived at the age when social acceptance is a matter of life and death. Girls made unpleasant remarks about the ugly blotches on his face; boys were equally unkind. Timmy hid from schoolmates, turned into a shy and severely introverted boy.

To make matters worse, the acne condition was so severe, the fistules would open up, necessitating medication. On one such occasion, he was excused from class and told to visit the school doctor.

"I wash and scrub my face all the time," declared Timmy, trying manfully to keep his upper lip from trembling before the examining doctor who was the first person the boy could confide in about this embarrassing problem. "I put creams and ointments on my face at night . . . but the pimples keep coming back."

The doctor noted the blotches and ridges and the glossy appearance.

No doubt, diet was largely responsible, but to what extent. Then he asked, "What did you eat for lunch today?"

"My mother gave me a fried egg sandwich, a piece of apple pie, then I had a soda from the vending machine."

"Do you remember what you ate last night?" asked the doctor.

Timmy answered brightly, "Sure. Spiced cold ham, mashed potatoes, a baked apple with whipped cream and a glass of milk."

The doctor noted that Timmy's nourishment, if it could be called that, came from processed foods. He ate meats and fish foods that were either pre-cooked, or processed so that they could be eaten with a minimum of effort.

Further tests showed that he was woefully deficient in the B-complex vitamins needed for a clear skin and healthy nervous system, the valuable Vitamin A that was a must for internal skin nourishment, and also deficient in *trypsin*—the enzyme which has the job of breaking down protein foods to release valuable amino acids. No doubt, Timmy's entire enzymatic system was out of kilter. Furthermore, *prepared and overcooked foods had reduced his chewing to a minimum*. This meant that his digestive tract did not secrete the flow of juices needed to attack foods, to extract their valuable nutrients. He was, to put it bluntly, "enzyme sluggish."

The doctor had a long talk with Timmy's mother, urging her to change her methods of cookery. The first rule was to eliminate as many prepared foods as possible. The second rule was to serve as many fresh plant foods as she could find—to be eaten with each and every meal. Spices and foods containing condiments were to be eliminated because they hindered function of enzymes.

Timmy was then told to drink fresh fruit and vegetable juices—prepared by a juice extractor by his mother, at home. Timmy was then told to eat as many natural foods as possible, as uncooked as palatable, and to chew . . . chew . . . chew. . . . .

The idea in prescribing chewable foods was to cut down on cooking. Except for meats, fish and eggs, and certain vegetables, this meant Timmy would have to chew much more than always. Thusly, his digestive flow was improved. Enzymes poured from the entire digestive tract, extracting nutrients from foods eaten to be sent all over Timmy's body via the bloodstream to nourish his mind and his physical condition.

Three weeks later, after changing to a raw and natural food diet, his acne condition began to subside. Gradually, throughout the fourth week, his skin became healthfully smoother and more radiant.

Enzymes were performing this miracle change.

Shortly thereafter, he was elected Class Leader and was much sought after for dances and social affairs.

Use stone ground 100% whole wheat flour.

Because so much cooking and eating calls for flour, it is important to discover one type of flour that is especially favorable for enzymatic function. At one time, back in the days when flour was made by means of the donkey mill, treadmill, windmill, water mill, energy was thus harnessed to turn the millstones. There was no need for any form of preservative or harsh chemicals to keep flour in storage for long periods of time.

Today, shipping, storage, shelf time, all call for a multitude of processes and ingredients that are destructive to enzymes as well as valuable nutrients.

So, I am happy to inform you that you can still obtain natural whole wheat flour, every bit as good as in the old days. The health food stores carry baked goods made with this flour, as well as the flour itself for your own baking. Why is this so superior?

All grain has a "germ" or embryo section. Wheat germ, consisting of about 2% of the wheat kernel, is a prime source of topnotch protein and enzymes—as well as those vitamins such as B-complex and E, which work together to spark enzymes into function. The oil, from which the grain's flavor is derived, comes from the germ section.

Modern processing calls for the flour to be de-germed. If the germ stays in the flour, it is perishable. So, to prolong its "shelf-life," the germ has to be removed . . . and preservatives added to give the flour even more keeping-time.

Without the germ, there is no Vitamin E. Without Vitamin E, enzymes are robbed of a valuable spark-plug. Small wonder that stone ground 100% whole wheat flour is gaining speedy popularity.

In stone grinding, the cool, slow rubbing between thick granite stones will evenly rub the germ oil throughout the flour. When the oils are evenly spread through the flour, there is no concentration of oily flakes that will oxidize and become rancid.

This is different from high speed milling methods which isolate the valuable germ, cause it to become rancid and spoil, hence its removal.

When the germ oil is slowly rubbed between cool stones, the germ so fully fuses with the flour, no concentration of oils remain. Stone ground flour has flaky particles which create a unique bread texture you will not find in other flour. Furthermore, a cool, slow milling process of the entire wheat berry retains precious vitamins and minerals that are lost in processes of high speed roller milling and separating. These are the same nutrients which work in harmony with enzymes to provide the life giving body processes that you must have to feel young, keep young, look young.

Health food stores as well as gourmet diet houses have bread products made of this stone ground whole wheat flour . . . and the flour, itself, for your own baking.

A SUMMARY OF HOW TO SHUN ENZYME ENEMIES IN CHAPTER 5.

1. Enzymes in foods are destroyed by cooking and processing. Safe enzyme temperatures are between 32° F. (freezing) and 104° F.

2. Enzymes are destroyed when water is boiled at 212° F. so remember how cooking can result in enzyme-less foods.

3. Raw foods keep test animals looking young and vigorous—they also keep humans in the prime of life.

4. Cook *what* must be cooked. Eat raw what *can* be eaten raw.

6

HOW CHEWING CAN REJUVENATE YOUR DIGESTIVE SYSTEM

Chewing stimulates enzyme action.

For enzymes to act out their role in digestive functions, you need to chew. Digestion is possible *only* when you have enough enzymes to break down into assimilation-form such essential foods as fat (35% to 45% of average diet), starch (35% to 50% of average diet), protein (12% to 15% of average diet). As you chew, a chain reaction is set up whereby enzymes become alerted and made ready for the food that is being swallowed. Chewing alerts and stimulates these enzymes, signalling them to get into action since they have work to do. As stated earlier, good digestion begins in your mouth.

Protein foods MUST be chewed. Dr. Lelord Kordel, author of *Eat and Grow Younger,* explains how properly chewed protein foods are welcomed by your system. "Digesting protein food is the main job of your stomach (carbohydrates and fats, except milk fat, are processed in the upper intestinal tract.) And to break down the proteins you eat into the amino acids which are the only forms of protein that can be carried by your bloodstream, Nature has provided your stomach with digestive juices that normally are strong acids. These acid juices, when strong enough, reduce all protein foods to a pulpy mass. The fibers of that steak you eat, when properly digested, should become as liquefied as though the meat had been pulverized in a powerful grinder and mixed with water."

How are these acid juices—these enzymes—made strong enough? By the process of chewing. Those of you who wail that your stomach is not as good as it used to be are cheating yourselves. Your stomach can be rejuvenated and restored to full digestive power—if the enzymes that "rule" your stomach are given an opportunity to attack foods. Chewing will call them forth into action.

Dr. Kordel also adds that not only is the hydrochloric acid enzyme in your stomach juices an effective pulverizer, it is also a powerful germ killer.

Enzymes kill germs.

"When your stomach is healthy," says Dr. Kordel, "that is, when its juices are strongly acid—your food digests instead of putrefying. Ordinarily, the billions of bacteria and fungi that get into your stomach each day fall into a strong acid bath and are destroyed. But if your stomach acids are not potent enough to kill these invaders, or to do a thorough job of pulverizing your protein foods, you may be headed for trouble—the kind of trouble that makes you feel ill, dispirited and too 'old' to get the most out of life."

Begin with tasty foods.

Mouth and stomach enzymes are strongly influenced into action by your desire for food; the sight, smell and taste of delicious edibles (not to mention a calm emotional state) all vitally control the quantity of enzymatic flow. To start a meal with disliked foods is destruction to enzymes. You reject these foods, mentally and physically, and you gulp them down to "get them out of the way." This is a ruinous eating habit. It is a step toward illness.

Try this enzyme stimulant.

Before you start a meal, drink this Enzyme Stimulant—in a large glass, mix soya milk with two tablespoons of powdered skim milk, one tablespoon of buckwheat honey and four tablespoons of wheat germ flakes. Stir vigorously or use a blender. Sip slowly. This Enzyme Stimulant will serve to awaken sluggish digestive fluids and prepare them for their job of working on foods to be eaten. The *first foods* of every meal should be your favorite foods.

The glands in your stomach (as well as the salivary glands in your mouth) are stimulated into action by desire for food; by the sight, smell and taste of appetizing food; and by the *calmness* of your emotional state. Your thoughts, and the correlative emotions produced have a great influence on your digestive juices.

Disturbing or irritating thoughts are all it takes to jam your digestive machinery. Bad news, angry words, nauseating subjects or seeing unappetizing food prevent you from satisfactorily digesting your meal.

Chew for acid distress.

Proper chewing, proper selection of foods, proper environment for eating will give your stomach a healthy and youthful enzymatic flow. This means you can discard those "alkalizer" powders and tablets which do little more than cause you to burp a bit; at the same time, they succeed in reducing the acid amount in your stomach—and you already might have *too little acid* which is responsible for your indigestion. Chewing stimulates the flow of hydrochloric acid by *natural* means.

Fat and thin stenographers.

Tension was high in the office. Long columns of figures had to be tabulated on the comptometer. Dictation was technical and boring. Two stenographers worked long and hard hours. They lunched at the company cafeteria because time was of the essence. This meant they were surrounded by other employees who talked of nothing but business. The two girls could not get away from the scene of tension.

Jean was thin. That is, at 24, she had a slim figure, with curves in the right places. Her co-worker, Clara, was pleasingly plump. She evoked smiles from those who watched her walk, with unsightly globs of fat bouncing up and down. Being plump was not pleasing to Clara.

"I don't understand it, Jean," she lamented to her friend as the two of them were eating the same lunch; it consisted of a healthy raw vegetable salad, a protein-rich tuna fish sandwich on pumpernickle bread, side dish of fruit flavored gelatin and a cup of yogurt. "We eat the same almost every day. Even at home, I've followed your diet plan. I've cut out sweets and pastries. But I'm fat . . . you're slim. What's your secret?"

Jean looked at Clara's plate. "You're finished with your sandwich, I see. And now you're halfway through with the gelatin." Jean pointed to her tray. "See? I'm still munching on my vegetables. There's plenty of time."

"What does that have to do with getting fat or slim?" wondered Clara.

"I *chew* my foods, giving them a thorough mouth digestion, so to speak, and when they reach my digestive system, they're less fattening."

Clara listened to the way Jean arranged her eating schedules. She would not sit down to eat if she did not have the time to properly chew all of her foods. This was her "slimming secret." It took three weeks before Clara, after following her advise, declared, "I've lost two pounds!" Within ten weeks, she was showing curvaceous lines.

Chewing de-starches those starches.

All fruits, vegetables, starch-sugar edibles *must* be thoroughly chewed. Carbohydrates cause weight increase. Carbohydrates will add unsightly and unnecessary pounds if they are not properly assimilated. They are nullified if they are digested in an alkaline base. Their fat-producing powers are thereby cut down and often eliminated. Your mouth saliva is an alkali and comes forth from your salivary glands. This is the first step in the long digestion of carbohydrates.

Suppose you bolt down your starch foods—these include some vegetables as well as all bread products such as rolls, cakes, pastries, macaroni, spaghetti, etc. By quick swallowing without proper chewing, the starch cannot be "treated" by your mouth enzymes and their alkaline actions. The starches reach your upper intestinal tract where an incomplete processing occurs and then are sent to storage depots of your body—to create fat and gain weight.

When the mouth enzymes which are alkaline are used to really digest starch *during the chewing process,* a great deal of their excess weight producing factors are made weaker. Ever watch herbivorous animals—those non-meat eating creatures such as the horse, cow, rabbit, etc? Watch them eat. They chew and chew and chew. They are obeying the law of Nature—thorough mastication of carbohydrates so that when the mouth enzymes (alkaline) mix with them, they are then ready to be assimilated by the system.

Fletcherism—put teeth in your stomach.

A few generations ago, Horace Fletcher, a nutritionist, developed a complete dietetical healing system in itself with enormous success for himself and others. His theory was to eat whatever kind of food you desired, but chew every bite completely. He also said to chew each bite 30 times. You could eat anything, so said Horace Fletcher, provided you obey the chewing principle—promptly dubbed Fletcherism.

This was going a bit too far. There is no reason to count your chews. What happens if you chew 31 times? Or if you chew 29 times? But

Fletcher was on the right track because it taught people the value of preparing a proper enzymatic flow by chewing and many who could not eat were cured by proper chewing.

The disadvantage of excessive chewing is that you reduce foods to a liquefied mass while still in your mouth and you may defeat your purposes that way; in effect, you are on a liquid diet! Chew . . . but don't overdo it! Chew until the food is in a comfortable *bulk* mass that can be swallowed so your enzymatic system can complete the job.

Eat small meals.

Your enzymes are happy when they have work to do—*but not overwork!* You will be more sensible to eat five or six meals in the waking hours—and preserve your health, strength and youth—than to stuff yourself routinely, three meals a day. Whoever said it was compulsory for you to eat three times daily?

The stomach that becomes overloaded on regular schedule, three times daily, gets old before the rest of the body. The stomach that is given smaller supplies of food every three hours is one that will be grateful for such kind treatment and keep your system healthy and young. And . . . by smaller meals, I do *not* mean those atrocious snacks of candies and sweets which should be an exception, rather than a rule.

Mary garden's secret.

How did this famed operatic beauty remain young, vivacious, alert, with sparkling eyes, even in her seventies? She told an interviewing reporter who commented on her youthful figure and energic disposition, "I never ate those eight and ten-course dinners. I have never in my life overeaten! I like everything inside the animal. Liver, tongue, sweetbreads, I adore. And vegetables out of the garden—but not out of the can. Bread I never touched—I was far ahead of my time."

These few words sum up the secret of instant health—enzymes keep you young and healthy. Small meals give your enzymes a chance to work with comfort, not exhaustion.

Chew all foods. With the exception of liquids, you must chew all foods for enzyme health and power. Dr. Leslie O. Korth, in *Some Unusual Healing Methods,* offers this advice: "In passing, it may be mentioned that starches and ordinary sugars are, or should be, largely predigested in the mouth by the saliva. The more we chew, therefore, the better can the saliva (via its enzymes) do its work before the food is swallowed.

"Protein, on the other hand, is digested in the stomach by the action of hydrochloric acid. Why need we bother, then, about chewing meat, for example? It will be obvious that by biting into the meat we break it down into particles, and the tinier these particles become under the grinding action of the teeth, the more surfaces will be exposed to the action of the gastric juices when the food reaches the stomach. Also, the mixing of the saliva with the meat particles 'liquefies' the bolus, or mass of masticated food, so that swallowing is greatly facilitated; this, of course, applies to all kinds of foods."

Make the most of tension.

Tension can be good for you! That's right. *Tension can make you healthy and improve your digestive powers.* Does this sound contrary to everything you have read and heard? Well, this is the truth. Of course, I don't mean the nervous tension that turns your stomach, churning painful wrenches so that eating under such circumstances only adds to the distress. Your attitude toward tension is the determining factor.

Earlier, you read of plump Clara who did not eat fattening foods, yet kept putting on weight. Why? Because she was so nervous and tense, she had no patience to chew her foods. She gulped them down. She would have been wise to self-hypnotize herself by withdrawing from her surroundings while eating . . . then chewing.

How can tension help you to chew? Consider this hint from Dr. Korth: "There is a great satisfaction to be derived from getting your teeth into things, both actually and figuratively. So bite . . . bite . . . bite your food; keep on biting hard with the full consciousness focussed upon the act. Not only will you arise from each meal with an easy, satisfied feeling in your body, and experience a clarity of mind, but you will be more or less freed from noxious aggression, especially when things go wrong."

Chew and think better. You may effectively lessen impatience, confused and illogical thinking by this little exercise: don't put any more food into your mouth until the preceding supply has been chewed to a pulp.

Home guide for stomach-enzyme rejuvenation.

Here is your working plan designed to improve the functions of your stomach-digestive and enzymatic system:

1. *Eat only when hungry.* Supply the demands of your body and mind.

not habit or custom. A healthy and natural hunger is the "voice of nature" telling us that food is required. There is no other true guide as when to eat. The time clock, schedule or noon whistle is not a true guide. Of course, you cannot walk off a job and eat. But you can train yourself so that you will be hungry during certain scheduled eating times when you have freedom. But if you come home at night, precisely at six thirty, that is no reason for you to immediately sit down and eat. Genuine hunger is a mouth and throat sensation and depends upon an actual physiological need for food; muscular stomach contractions accompany hunger and are thought by physiologists to give rise to the hunger sensation. Give yourself time.

2. *Avoid eating when in pain, fever, mental and physical discomfort.* You have heard of *anorexia*—a loss of appetite accompanying such conditions as great fatigue, chronic disease, grief, anger, hysteria and conditions of mental upset inhibit the flow of enzymes. To thrust food into a stomach under such circumstances is to invite disaster! You invariably feel so-called gastric or intestinal distress. Anorexia further inhibits enzymatic flow. Nature wants you to rest until your health is restored. Should you eat during a fever? *No!* The fact that a coated tongue, which prevents the normal appreciation of the flavors of food, prevents the establishment of gustatory reflexes and, through these, the secretion of enzymes, should show the values of enjoying your food or not eating at all. The feverish person needs a fast, not a feast. Clear soups, bland edibles that require a minimum of enzymatic action are the best in such situations.

3. *Avoid eating during or immediately after or before work or heavy mental and physical effort.* The Romans had a proverb: "A full stomach does not like to think." You need leisure time to eat and digest foods. *Eat only when you are able to then have a bit of leisure time to digest.* After a hearty meal, animals retire to a quiet resting place to give their enzymes a peaceful environment in which to work; they rest while the enzymes work happily.

Good enzymatic power requires that almost the entire attention of the system be given to the task. Blood is rushed to your digestive organs in large quantities. There is a dilatation of your blood vessels in these organs to accommodate the extra supply of blood. There must be an harmonious constriction of the blood vessels in other parts of your body in order to force the blood into these digestive organs to compensate for their own loss of blood.

Digestive power must have much energy and valuable blood power. The comparative lassitude which follows a hearty meal is proof that this supply of energy and blood force is at the expense of the rest of your body. To force yourself into physical or mental activity either before or immediately after a meal is to deprive your digestive system of this needed vigor.

Hearty eating when you are tired from either mental or physical work is likely to be followed by indigestion, malaise and incapacity for work. Rest and sleep is needed, more than food. After relaxation, then you may eat.

4. *Do not drink with meals.* This is most important—by drinking is meant *any* liquid beverage. (Animals and so-called primitive peoples do not drink with their meals and there is every reason to consider this instinctive practice to be best.)

Water leaves your stomach in about fifteen minutes after being ingested. Water carries the diluted and consequently weakened enzymatic juices along with it. This interferes seriously with digestion. Some maintain that water, while eating, stimulates the enzymatic flow. What is the answer? This is not a natural way to spark your enzymes; it is of no value to your enzymes to be secreted only to be seized up by water and carried out of your stomach into your intestines before these enzymes have had time to act upon food.

Water may be taken about two hours after you have finished eating. At this time, the reactions of enzymes are well under way, if not fulfilled, and water helps to sweep these enzymes onto your intestines. Drink water about thirty minutes *before* a meal. You may drink water thirty minutes *after* a fruit meal; two hours *after* a starch or carbohydrate meal; four hours *after* a heavier protein meal.

Drinking with meals leads to the bolting habit. Instead of thoroughly masticating and insalivating your food, if you drink while eating, you are tempted to wash down foods half-chewed. Milk is regarded as a food. Sip slowly and hold in the mouth until enzymes have attacked it, then swallow. Take no other food in your mouth with milk.

5. *Avoid cold drinks.* Ice cold drinks are enemies of enzymes. This includes iced water, lemonade, punch, iced coffee, tea, etc. Cold will stop the action of the enzymes which now must wait until the stomach

temperature has been raised to normal before they can resume their digestive activities. A cold drink, introduced to your stomach, will shock and chill it . . . and shock and chill your enzymatic flow. After the cold water is sent out of the stomach, a reaction sets in. You feel feverish and very thirsty. Ice cream acts in the same way. Eating ice cream is like putting an ice pack to your stomach.

6. *Avoid hot drinks.* Hot drinks weaken and enervate digestive enzymes. These destroy the tone of digestive tissues, weakening the power to act mechanically upon the food. The weakening of its tissues in this way may lead to prolapsus of the stomach (unkindly called pot belly). Enzymes are hindered in extremes of hot and cold. The functional powers of enzymes are at their highest when working in a temperature conforming to that of the normal body temperature—or at least when the temperature is not over 100° F. *Room temperature* is a good guide. Or, at least, liquids should be consumed after the chill has been taken out by "thawing" at room temperature; or, after the heat has been reduced so the beverage can be sipped slowly with no discomfort.

7. *Eat small meals.* If your enzymes are to work at normal level, almost the entire attention of the system must be given to the work. Digestion, as explained above, calls for a rush of blood, dilatation of blood vessels, etc. At the same time, your brain and muscles want to work, too. They also need blood. If one part (your digestive system) gets an extra supply, some other part must get less (your brain, muscle and nerve system). So eat smaller meals to give all of your system a sufficient amount of blood.

Be Kind to Your Enzymes

Jonathan Forman, M.D., editor of the Ohio State Medical Journal, urges that we be kind to enzymes if we want to reap the benefits of health, happiness and long life. "To carry on the life processes, each of these (billions of) body cells must digest food, excrete its wastes, repair itself and be ready in one-millionth part of a second to perform its specialized task.

"Every reaction depends upon a chemical response triggered and controlled by a specific enzyme system. All life is merely a great number of these enzymes working together in coordination. In other words, life is the function of the protein of the cell joining with a particular vitamin and one of the mineral nutrients originally from the soil, to form an enzyme with a highly specialized function."

KEY POINTS IN CHAPTER 6.

1. You can stimulate enzyme flow by chewing foods—chew very thoroughly.

2. Boost your appetite with the Enzyme Stimulant drink.

3. Overcome acid distress or excess acid indigestion by chewing while you remain calm and relaxed.

4. De-starch foods by chewing and remain slim even though your foods may be carbohydrate-rich.

5. Eat small meals.

6. For dry mouth, chew until your enzymes come to the rescue with a natural lubrication via salivary flow.

7. Chewing can ease tension, make you think better.

8. Eat when you are hungry; shun eating when in pain, fever, mental and physical discomfort; avoid eating during or immediately after or before work or when faced with heavy mental and physical effort. Eat when you have enough leisure time to give your enzymes a chance to digest foods.

9. Do not drink water with meals; wait two hours after eating before taking an average large sized beverage. Avoid extremely hot or extremely cold drinks.

7

FASTING—
MIRACLE HEALTH BUILDER

How would you like a brand new body and an alert mind? Picture a happy life, free from disease and worry, untouched by the ravages of time and the pace of modern civilization. Is this just an illusion or can it happen to you? .

You *can* rebuild your body and rejuvenate your mind. You *can* restore the youthful health of your digestive system. You *can* super-charge sluggish enzymes so they will give you a clear skin, a sparkle in your eyes, a life to your steps. How? *By fasting!*

What is a fast?

The word fast is derived from the Anglo-Saxon word, *faest*, meaning "firm" or "fixed." The practice of going without food at certain times was called *fasting*, from the Anglo-Saxon, fasten, to hold oneself from food. Like most English words, fasting has more than one meaning. A dictionary defines it as "abstinence from food, partial or total, or from proscribed kinds of foods." Insofar as religious fasts are concerned, abstinence from proscribed foods is what this means.

We may define it as thus: *Fasting is abstention, entirely or in part, and for longer or shorter periods of time, from food and drink or from food alone.*

Specialized fasts. You may have heard of fasting with respect to particular diets. For example, a fruit fast is one in which you abstain from fruit. A milk fast is abstinence from milk. A water fast is abstinence from water, and so forth. These are very limited fasts and may be advisable when you are allergic to certain types of foods. But such fasts would not do much to rebuild your health since you still continue to eat as much as before, thereby giving your digestive system the same work as always.

Fasting can cure colds.

Although the poorly fed body is apt to catch cold and suffer from various infections of the respiratory tract and mucous membranes, a fasting person seems seldom to catch cold.

Salesman cures own cold.

"I travel over 2500 miles a month, going from one climate to another, but I don't catch cold," boasted a young salesman on a train headed north into the cold Canadian climate. His companions were already sniffling, sneezing, using boxes of tissues.

"Well, guess you were just born lucky," sighed another salesman, reaching into his overnight bag to withdraw a warm shawl that his wife gave him. "See these bottles?" He pointed to an array in the bag. "I take one pill here, one pill there, and it keeps my cold under control."

"But I don't take anything," the cold-free salesman declared. "I dress warmly. I take care of myself. And as soon as I get any signs of a cold—I stop eating."

The others stared at him. One was aghast. "You mean you starve yourself? Is that your secret? I thought a sick person needs strength and has to keep on eating. In fact, I eat more when I feel sick."

"That's why you keep catching colds. Nature sends a sniffle as a warning signal that you're body is overloaded. Now, I don't mean you have to take laxatives. You need a more natural internal cleansing—and fasting will do it. By fasting, I don't mean starving. Yes, I eat during my fast—but I eat as many as eight small meals a day—and my foods are light. Very light. My protein is from the soybean. My vitamins and minerals come from a raw vegetable source. My amino acids are from the more easily digested foods. Fats come from fish, rather than meats. I give my body a chance to cleanse itself and get rid of the start of a cold." The healthy salesman knocked wood. "Lucky me? I suppose so. But fasting does the trick."

Nature's fast-rest cure.

"One of the best remedies which is always at our disposal is fasting," states Dr. Alfred Vogel in *The Nature Doctor.* "If we do not feel well because we have eaten too much, if our stomach is disturbed because we have eaten some unsuitable or even bad food and we are suffering from sickness and diarrhea, then fasting is the most natural remedy. Our domestic animals behave more sensibly than we do in such situations: they refuse all food! Dogs and cats usually eat grass, so as to bring up the mucus, and follow this with at least one day's fast. Animals know instinctively that they must not eat until they feel quite well again. These natural health regulations are followed even more closely by wild animals: they, too, know of no special remedy beyond fasting, if they are ill. They just lie down in the shade and rest and fast until they feel well again. When we fast, the body gets an opportunity to rid itself of harmful metabolic accumulations."

Managing a rational fast.

By rational it is meant that you do not suddenly retire from the world and actually starve yourself. A rational fast gives you enough nourishment to go about your daily job—you eat, but do not stuff. Dr. Vogel says, "From time to time, you should arrange to have a fruit juice day and follow this by two days taking only water. Before you begin a fast, see that the bowels are empty. Linseeds or a herbal laxative will help you to do this. (Herbal laxatives are sold in nearly all health food shoppes all over the country.) If your liver is in good order, you can now begin the fast either with orange-grape-fruit or grape juice; or during the berry season, you could make use of berry juices.

"Any waste left in the body will thus be disposed of and your organs will begin to function better. Should you begin to feel sick during the fast, you would be well advised to speed up elimination by encouraging skin-function: cold friction baths will open the pores of the skin and stimulate circulation, while deep breathing will help likewise. Short walks in the woods and along the country lanes, will soon restore a feeling of health and well-being again."

Enzymes rest during fast.

Your enzymes are constantly at work. They need rest just as do other body organs such as your muscles, heart, eyes, ears, lungs, etc. Enzymes work hardest when attacking meat protein foods. They have slightly less work with fibrous fruits and vegetables. Their work is still eased when working upon dairy foods and cheeses. During a fast, restrict your meat intake to a minimum. Select more digestible meats such as baked lamb, broiled chicken, broiled chops—*very lean!* Trim off *all* visible fats.

Eat more liberally of steamed and baked fish foods; eat nominal amounts of baked chicken. As for dairy foods, select natural cheeses—cottage cheese, farmer cheese, ricotta, yogurt as a milk food.

Fruits and vegetables should emphasize bananas, figs, dates, prunes, melons, all citrus fruits, grapes, tomatoes; also Brussells sprouts, cabbage, cauliflower, lettuce, mushrooms, peas, squash. Eat them raw, wherever possible. And remember: *Chew* to a liquefied pulp to spare the work on your enzymes and give them a chance to rest.

Try meat substitutes.

Few people, surprisingly enough, are aware of meatless foods. That is, especially prepared foods made of vegetables and gluten (cohesive substance in wheat) and soybeans. These are prepared to taste like chicken, fish, chops, steak, meat loaf, scallops, liver, etc. Health food stores sell them in canned form. Brand names include Worthington Foods, Loma Linda, Emenel Foods, Lange, Elam, to name a few. To eat them, you could hardly believe they are made from vegetable or peanut stock. Some have a seed or nut base. Others are made exclusively from vegetables. None are made from meats. Even frankfurters (made from carrot base) are deliciously deceptive. To break your meat habit during your fast, serve these meat-substitutes. They are prepared just as are meats. Follow directions on the label. Most health food shoppes have a wide variety of such foods.

Ease hunger.

If you become hungry, chew a few raisins slowly and thoroughly. Desert Bedouins live for days on end on nothing but a few dates. They derive maximum benefits from such a humble meal by chewing . . . chewing . . . chewing, thereby predigesting such foods properly. Who ever heard of a Bedouin with a stomach ache from over eating? You can help yourself, too, by chewing on raisins or dates, making maximum benefit from the good grape sugar.

Case of mis-used fast.

Both Lillian and Arthur were overweight. In addition, they were subjected to periodic backaches, nervous tension, chronic headaches. Lillian developed sniffles upon the first gust of a breeze. Arthur's asthma made life miserable for both of them.

"Let's go on a fast," suggested Lillian, one evening after both had finished a sumptuous meal that made the table groan from the weight of the foods. "It's possible we're overloading ourselves and raising our temperatures to the point where we're sensitive to everything."

Arthur was reluctant. "I like to eat. I don't want to starve. Besides, I have to go to work every day. If I starve, how can I get the strength to work?"

"We're going to keep right on eating," explained Lillian, but no more heavy foods. We're cutting out meat-eating every day. Weekends, we're going to eat raw fruits and vegetables—and nothing else, except some whole wheat bread with caffein-free coffee. During the week, we'll eat eggplant steak, carrot loaf, peanut stew, and so forth. And we're *not* going to stuff ourselves with those lighter foods, no matter how much more digestible they may be."

Lillian was receptive to the idea and followed a diet plan which called for abstention from meat five days of the week. On Tuesday and Thursday, she served broiled meat patties for dinner. Otherwise, they ate light courses of fish and meat substitutes.

Two weeks later, Lillian noticed her nerves were no longer taut. She had not suffered from a headache for three days. She watched television for hours at a time, yet suffered no blinding flashes. She lost two pounds. She went out without her sweater and the cool Autumn breeze invigorated her. It did not send her to bed with an attack of sniffles.

"This fast is curing me," exclaimed Lillian.

"It's making me sick!" declared Arthur. "I'm pale. I'm weak. I snap at the office boy. I growl at my secretary. I'm swearing off this silly idea. In fact, I never wanted to do it anyway."

He returned to his heavy eating. Lillian continued on her rational or controlled fast. Gradually, she increased her quantities but never again touched gravies, rich pastries, heavy portions of heavy foods. Only after she was completely self-cured of all her ailments, did it make an impression on Arthur.

"Why should it work for me, not for you?" asked Lillian. "Because you didn't cooperate. Also, you can't relax. Tension worked against your fast. In another three weeks, we're going to Florida. Two months of sunshine and relaxation, Arthur. Then you're going to be more receptive to a fast."

Yes . . . under tranquil conditions, a fast did wonders for Arthur and he returned home, a brand new person.

Ancients knew of fasting.

"Instead of using medicine, rather fast a day." So spoke Plutarch, the Greek writer at the time of the First Century. A contemporary of his, Aurelius Cornelius Celsus, a Roman medical writer and Epicurean philosopher credited with writing *The True Discourse*, declared, "In extreme distress, a sick man should take nothing. In a second degree (not too serious an ailment), he takes nothing but what he ought."

Religious fasting. Fasting as a religious observance, has long been practiced for the accomplishment of certain aims. Religious fasting is of such early origin, it precedes recorded history. Partial or entire abstinence from food, or from certain kinds of foods, at stated seasons, prevailed in Assyria, Persia, Babylon, Scythia, Greece, Rome, India, Ninevah, Palestine, China, in northern Europe among the Druids, and in America among our own Indians. It was a widely diffused practice, often indulged as a means of penitence in mourning and as a preparation for participation in religious rites such as baptism and communion.

The Ancient Mysteries, a secret sect at the dawn of civilization, that flourished for hundreds of centuries in Egypt, India, Persia, Thrace, Scandinavia, the Gothic and Celtic nations, prescribed and practiced fasting. The Druid religion among the Celtic peoples also called for a long probationary period of fasting and prayer before one could be accepted into the faith.

A fifty day fast was required in the Persian Mithraic religion. Indeed, fasting was common to all sects of the time. Moses, who fasted for more than 120 days on Mount Sinai, came down and professed to have learned "all the wisdom of Egypt" because of this abstinence.

The American Indians fasted in acquiring their private Totem, in the belief that prolonged hunger gives rise to wisdom.

Zoroastrianism is the only religion which forbids fasting, for some undisclosed reason.

Bible and fasting.

Fasting was practiced by nations and individuals throughout Bible times. The most important fasts recorded in the Bible are: Moses before approaching God on Mount Sinai to receive the Ten Commandments (Exodus 24:18, 34:28); Elijah for 40 days before reaching the mount of

God (I Kings 19:8); David for 7 days when his child was ill (II Samuel 12:20); Jesus for 40 days (Matthew 4:2); Luke declared: "I fast twice in the week." (Luke 18:12); "This kind cometh not out except by prayer and fasting." (Matthew 17:21); a fast declared throughout Judea (II Chronicles 20:8).

The law of Moses ordered a yearly fast on the Day of Atonement or Yom Kippur, the so-called Jewish New Year.

The Bible cautions against fasting for mere notoriety or selfish gains (Matthew 6:17,18). It also advises those who fast not to be sad and gloomy (Matthew 6:16) but to find pleasure in fasting and to perform one's work (Isaiah 58:3). Also, certain fasts shall be fasts of gladness (Zecharias 8:19).

The mysteries of Tyre, which were represented in Judea in the days of Jesus, in a secret society known as the Essenes, also prescribed fasting. In the first century A.D., there existed in Alexandria, Egypt an ascetic sect called Therapeutae, who borrowed much from the mystical Kabala as well as the Pythagorean and the Orphic systems. These Therapeutae gave great attention to the sick and maintained that fasting was a *curative* measure.

Mahatma Gandhi who was assassinated at 84 years of age, fasted many times; often, up to 50 days. Gandhi, of course, fasted not only for health, but for moral and political considerations. One of his last fasts before his assassination lasted 42 days and was an attempt to reconcile Hindus and Moslems.

The writer of *Peregrinato Silviae*, in describing the observance of Lent in Jerusalem, in about 386 A.D., states: "They abstained entirely from all food during Lent, except on Saturdays and Sundays. They took a meal about midday on Sunday, and after that they took nothing until Saturday morning. This was their rule through Lent."

When we read that the Archangel Michael appeared to a certain holy man of Sipponte after the latter had spent a whole year in fasting, *we do not think of him as having abstained from all food, but from certain proscribed foods.*

This illustrates the marvelous wisdom of the ancients who knew how to rest a digestive system—by fasting and not necessarily starving.

Wrong types of fasting.

The practice of fasting until sundown, then feasting in the so-called fast of Ramadan, a practice of the Mohammedans, is not a proper method.

During this period these people do not eat. But as soon as the sun goes down, they start stuffing themselves. A visit to any Mohammedan city will show a grand festival at night, to make up for their day of fasting. The city holds carnivals, the restaurants are well illuminated, food is in abundance. The wealthy hold sumptuous banquets. At the end of this month long observance, they celebrate Bairam, or a special feast.

To deny yourself food all day, then engorge yourself at night, is not a suitable fast. It is just like splashing gasoline on a dying fire. Everything explodes.

Self-purification fasts.

In India, many undergo self-purification fasts. The leader of the Indian Socialist Party, Jayaprakash Narayan, fasted for 21 days so that he would be better able to fulfill his public tasks. However, he underwent this purification fast in a special nature cure clinic under the supervision of the man who watched Gandhi's health during his fasts.

Fasting as magic.

Tribal fasts, as seen among the American Indians, to avert some threatened calamity is the use of fasting as magic. Fasting was often part of the rite of initiation into manhood and womanhood or for sacred acts among many tribes. Fathers of newborn children are required to fast among the Melanesians. If we carefully distinguish between magic fasting and protest fasting, as in hunger strikes, we may say that magic fasting is undergone to achieve some desired end outside the person of the faster.

Strength-mind-power through fasts.

Socrates and Plato, two of the greatest of the Greek philosophers and teachers, fasted regularly for a period of ten days at a time. Pythagoras, another of the Greek philosophers would also fast regularly, and before he took a test at the University of Alexandria, he fasted for forty days. He required his pupils to fast for forty days before they could enter his class. These ancients believed that fasting would invigorate physical and mental powers.

Hunger strikes.

As a means of arousing public sympathy, some persons go on hunger strikes. We know about Gandhi, of course, as a prime example. Then there is the hunger strike of McSwinney, an Irishman who shunned food for 74 days in a Cork prison as a protest against English rule over Ireland. His co-political hunger striker, Joseph Murphy, died on the 68th day of abstinence. Remember when some suffragettes would also go on strike as a means of calling attention to their demands for women's rights?

In 1961, Master Tara Singh, leader of the Sikhs, announced he would go on a hunger strike in his dramatic demand for a separate Sikh state in the Punjab. On the same day, 76-year-old Hogiraj Suryadey, ascetic and religious leader, began his own hunger fast in protest against the Sikh's having their own state. Both hunger strikes thus neutralized each other; although, by maintaining the status quo, Suryadey won the contest. You may laugh at such hunger protests, but a battle of this sort is less taxing on the resources of the people and results in less bloodshed than does an old-fashioned shooting revolution. Maybe hunger strikes can replace atom bombs!

Mark Twain on fasting.

Writing seriously in *My Debut as a Literary Person*, this noted author says, "A little starvation can really do more for the average sick man than can the best of medicines and the best of doctors. I do not mean a restricted diet, I mean total abstinence from food for one or two days. I speak from experience; starvation has been my cold and fever doctor for fifteen years, and has accomplished a cure in all instances." Mark Twain tells of talking to several shipwrecked sailors who were forced to exist for long periods without food and without water. The sailors told how some of them had developed abscesses and ate like animals—*before being shipwrecked*. The more they ate, the greater was their illnesses. Yet, after nearly three weeks of near-starvation bcause of shipwreck, they emerged feeling healthy and fit. They were weak, true, but diseases of the ordinary sort had left them. Mark Twain was a regular faster.

The process of autolysis.

In order to understand the benefits of fasting, you should have an awareness of the process of *autolysis*. The word is derived from the Greek, meaning "self-loosing." It is used in physiology to designate the process of digestion or disintegration of tissue by enzymes generated in the cells themselves. Autolysis is a process of self-digestion and absorption.

Enzymes accomplish this process. They are called *oxidases* and *peroxidases*.

In conditions of illness, enzymes will be called upon to digest the substances of the diseased cells and thereby cause these digestants to be expelled from the body. Without being digested, these diseased particles remain. *Autolysis* is the function of breaking down the diseased materials and preparing them to be excreted from the body.

Here's an example of how autolysis works: did you ever have a pus pimple or abscess on the surface of your skin? You were told that its poisonous or infectious materials had to be drained and disposed of. How was this accomplished? By the means of enzymes—which attacked the infectious bacteria, digested the germs and removed them via drainage.

Let's take a step backward. How did the abscess get *outside* the skin when it was undoubtedly caused by an internal condition? *Autolysis* was sparked by your enzymes which digested the infectious bacteria underneath your skin—forced them outside in the shape of a pus pimple and thusly enabled it to be drained and removed.

How can you get autolysis to work in cleansing your internal and external infections? *By fasting.* When supplies of food are exceedingly small or entirely cut off for a considerable period of time, the enzymes look around for other things to act upon. Seeing any accumulation of cells, such as in a pimple, tumor, cyst, abscess, etc., they attack these sources of work. They digest these cells and cause them to be excreted through the pores of your skin. Before this autolysis process can be accomplished, fasting is needed.

Secret of animal fasts.

The seriously wounded animal refuses to eat, as was stated earlier. It crawls away—yet its wound heals. Great quantities of blood are sent to the site of the wound. This represents a great quantity of "food" given to the injured wound. The fasting animal draws upon its reserves of food materials out of which to repair its torn, cut or broken tissues. These are first autolyzed by enzymes, then carried to the part of the body where they are needed. Your body not only distributes nutrients to all organs, it can re-distribute them: i.e., taking them from one part to another needed part. It has the wonderful ability to shift nutrients around. Autolysis makes this process possible. Without autolysis, wounds would never heal. One part of the body would be perfectly healthy . . . the other part is diseased or even dead.

Enzymes are needed to perform autolysis—and the power of enzymes is increased when you fast because these internal messengers of the body now turn attention to attacking, digesting, and expelling infectious wastes. For autolysis to work, you must force your enzymes to turn attention to infectious accumulations in your body. Deny food to your enzymes and force them to become cannibals and devour these infectious accumulations. *Deny food via the fast!*

Fasting and growths.

The pocess of autolysis may be used to cope with certain growths and tumors. These growths consist of blood, flesh and bone. Tumors are composed of tissues, the same kind of tissues as other body structures. These tumors may be susceptible to autolytic disintegration, just as are ordinary tissues; and during a fast, such tissues do undergo a process of dissolution and absorption. Fasting will reduce the amount of body fat, muscle size and also tend to reduce the size of the tumor or in some cases may cause it to disappear altogether.

During fasting, the accumulations of superfluous tissues are overhauled and analyzed by the enzymatic "system;" all available component parts are utilized in nourishing essential tissues. But—the refuse or diseased substances are bypassed!

The famed Bernarr Macfadden once stated, "My experience of fasting has shown me beyond all possible doubt that a foreign growth of any kind can be absorbed into the circulation by simply compelling the body to use every unnecessary element contained within it for food. When a foreign growth has become hardened, sometimes one long fast will not accomplish the result, but where they are soft, the fast will usually cause them to be absorbed."

There are limitations to the autolyzing process, of course. In the case of a very large tumor, the ailing and fasting person will experience uncontrollable hunger before the tissues can be autolyzed. Some tumors are so situated that they dam up the lymph stream; they continue enlarging (feeding upon the excess of lymph behind them) despite fasting. Autolysis cannot completely absorb such tumors; but they may reduce them in size so as not to constitute a menace. Fasting must be prolonged so the enzymes are forced to seek food elsewhere and cope with the tissues of the growth.

Once again, remember that fasting is not starving. The word starvation is derived from the Old English *steorfan*, meaning to die from lack of food. Mention fasting and you think of starving. Not so. Fasting is the elimination of proscribed foods and reduction of foods, in general.

Body improvement via fasting.

We have seen how fasting and autolysis is Nature's own method of ridding the body of diseased tissues, excess accumulations of waste and toxins. What happens in your body during your fast? At the start, you have a temporary increase in elimination. Other eliminations, in the past, consisted of wastes caused by the daily intake of more food than you needed. Now internal wastes—surplus waste accumulations—are being eliminated.

The blood and lymph become purified. Pent-up excretions are expelled from your body. Your nervous system and vital organs secure relief and rest. Bodily irritations that caused mental irritations now cease; indeed, you are being "made over." Ridding your cells and tissues and fluids of accumulated toxic wastes accounts for the maximum benefits derived from fasting. These benefits last until the toxins accumulate which illustrates the need for periodic fasts.

There is no known natural method that can equal fasting as a means of accelerating the processes of elimination. When food is withheld only a short time elapses before the organs of elimination increase their work of discarding accumulated waste products. Secretions begin a physiological house cleaning.

What will fasting do? It rests your vital organs. It halts the intake of foods which decompose in the intestine and may contribute to internal intoxication. It will empty your digestive tract and dispose of putrefactive bacteria. It gives your organs of elimination a chance to promote release of accumulated waste products. It re-establishes normal physiological chemistry and normal secretions. It promotes the breaking down and absorption of effusions, deposits, diseased tissues and abnormal growths. It restores a youthful condition of tissues and cells, thereby rejuvenating your body. It clears and strengthens the mind.

All this, and much, much more is accomplished because fasting compels the body to rely upon its internal resources, forces the tearing down (by autolysis) of growths, accumulations, infiltrations, overhauling them and excreting them.

How to Fast Intelligently

"Fasting? I'd rather die than stop eating." As the mother of two teen age boys and the wife of an insurance adjuster, Phyllis G. made the aforementioned startling statement. "You know," she told her other three bridge partners one afternoon over the card table, "I thought I'd be abl to lose weight, get rid of this stubborn rash, even improve my mind, b fasting. That's right. . . . I was told I would be more alert and coul

obey my reflexes if I would fast. Well, I admit I failed my driver's test four times because I can't obey traffic signals on the button—but even if I sleep late and it takes me an hour to get ready in the morning, I just couldn't starve myself. I tried fasting. For a whole day, I just drank water. It was agonizing. No more fasts for me."

Her bridge partner looked up. "Well, *I* tried fasting. Remember all those headaches I had? And that awful ringing in my ears? I tried fasting —three times I relented. The fourth time I did it a better way."

Then she told of her method which is a guide for all other would-be fasters to follow:

Good fasting rules.

1. Taper off gradually. Do not suddenly shock your system by depriving food. True, ancients may have halted and resumed eating without difficulties but many of them ate sparsely. Many more were fruit and vegetable eaters, exclusively. A lifetime of sparse eating, with little or no meat, makes fasting a simple method. Reduce all food portions to a *comfortable* amount.

2. To satisfy that "meat hunger lust," use meatless foods—described earlier. These are available in special diet shops and health food store outlets throughout the country.

3. Increase your raw vegetable intake. This gives you a feeling of fullness and hunger satisfaction.

4. Slowly, reduce *all* quantities of *all* foods. Eliminate meat for two consecutive days. Then for three consecutive days. Now resume meat for one day. Then go back to a non-meat diet.

5. Drink as many freshly squeezed fruit and vegetable juices as possible.

6. Eliminate *all* prepared desserts such as pastries, cakes, pies, etc. These are rich in carbohydrate and cause gas distress during a fast. Substitute *fresh fruit* as a dessert following a meal.

7. Slowly, reduce quantities again until your system is adjusted to less food. Keep portions at a minimum.

"Is it as simple as that?" exclaimed Phyllis. "Maybe I've been doing it wrong. I just stopped eating!" Indeed, she *was* wrong.

"After following that fasting plan," continued her friend, "all of my petty annoyances began to vanish and now I feel like a new person. Nothing irritates me . . . BRIDGE!"

"Ohhhh!" yelled Phyllis. "I lost!" She screwed up her face in a crying expression.

Her friend laughed. "Follow my fasting plan, and you'll laugh at losing, the way I did."

While You Fast

1. *Keep moderately active.* Your thoughts will be averted from food if you keep active. Find something to take up your time. If you work, so much the better.

2. *Fresh air.* Good ventilation is important. Air should be clean and fresh.

3. *Keep warm.* This is very valuable. Chilling will cause discomfort. Also, when you fast there is a lower metabolic rate so you are more vulnerable to colds. You feel cold at what may be an ordinary and comfortable temperature. To become chilled means you cause a rapid loss of your reserves. Warmth promotes comfort. A hot pad to your feet will prevent chilling. Warmth is comforting and relaxing.

4. *Take warm baths.* Avoid extremes of temperatures. Take a quick bath and dry off quickly. A too hot or too cold or too prolonged bath may prove enervating. A bath should be closer to your body temperature. A sponge bath might also be utilized.

When to Break a Fast

When your tongue is clean, your sleep is restful, your skin clear, your eyes bright, there is no pain, and you feel a healthy hunger—resume eating. *But you no longer stuff yourself!* You now understand the values of moderate, but wholesome, eating.

How to Break a Fast

Here are your plans for after-fast eating:

1. Increase all portions of raw fruits and vegetables.

2. Meat may be eaten thrice weekly—but in moderate portions that are gradually increased.

3. *Begin* each meal with either a fruit dish or a raw vegetable salad.

4. One hour before each meal, drink a glass of freshly squeezed fruit or vegetable juice.

5. Completely eliminate all harsh condiments and spices. Substitute salt substitutes which are sold at health stores or at corner drug stores, if you must have a flavoring agent other than the taste buds of your invigorated tongue.

6. More than ever before—*chew, chew, chew* all foods!

MAIN POINTS TO REMEMBER IN CHAPTER 7.

1. Enzymes need rest—just as you need rest. Learn about fasting. You eliminate, from your diet, heavier foods and subsist on a raw fruit and vegetable diet for a period of 3 days. This is less taxing on your enzymes.

2. Fasting enables your body to dispose of accumulations of waste substances by means of *autolysis*. Enzymes devour waste products when they do not have heavier protein foods to work upon. In other words, give enzymes a chance to dispose of waste substances.

3. Fasting can help cure colds, build immunities, restore your body balance.

4. Fasting should be voluntary and with a cheerful disposition as well as anticipation.

5. Here's a nutshell summation on how to fast: The first day, cut down meat. Eliminate sauces, gravies, desserts (except fresh fruit desserts). The second day, step up your raw fruit and vegetable intake, drink as many juices as possible. The third day, continue reducing meat and all other cooked foods—subsisting basically on raw foods. The fourth day, entirely raw foods! The fifth day, continue on with raw foods, i.e., fresh fruits, vegetables, nuts, seeds. The sixth and seventh days are the same. When you begin your new week, you should feel rejuvenated by this one week fast. Slowly increase your meat intake on the eighth day.

8

BREATHE THIS OXYGEN COCKTAIL
FOR ENZYME POWER

Do your enzymes breathe?

Surprisingly enough, yes! Of the three ingredients needed by enzymes (the first two are food and water), fresh, pure air is the most vital. You can go without food for as long as 40 days and your enzymes can still function to keep you alive and in reasonable recuperative health. You can live without water for as long as 7 days before your parched enzymes die of thirst. But—if you are denied air for the maximum of 9 minutes, your enzymes choke and die—and so do you!

How air keeps you alive.

During normal breathing, air, together with its component oxygen, is drawn into your pharynx through your nostrils. From there, air descends into your trachea or windpipe. The trachea subdivides into two tubes, known as the bronchi. Each one supplies one lung.

As you breathe, air passes down these bronchial tubes, through their branching passageways until it reaches tiny air sacs in the lung tissue which are terminals of the tiniest bronchial subdivisions.

Oxygen in the air sacs diffuses through their thin walls and those of adjacent capillaries into the blood; the breathed-in carbon dioxide that enters the blood joins with the hemoglobin (the substance or pigments that gives blood a red color) in the blood and is carried to each and every part of your body. In the capillaries of your body tissues, oxygen is released by your hemoglobin so it can pass into the fluids of those tissues, then go to the body cells and to your enzymes. All this while, your enzymes are waiting for this pure oxygen which it needs to be able to stay alive and function—to keep you alive and healthy!

The constant demand for air.

Each time you normally inhale and exhale, about 500 cubic centimeters of air (about 16 fluid ounces) come in and go out. During strenuous exercise, the amount may be as high as 4000 cubic centimeters (about 128 ounces). You breathe about 10 times each minute throughout your whole life. You breathe whether or not you are aware of it. Nature has seen fit to create this arrangement since your enzymes make such a constant demand for air, it must be an automatic process.

True, you can hold your breath (remember how you did this when a child?) but with the greatest effort, you cannot hold out longer than 40 seconds. Your body refuses to permit further deprivation. Nature has made the demand for air so insistent that enzymes just will not allow you to choke them to death, voluntarily. You can store food and water in your body and then go without these elements for days at a time—not so with air.

Threats to enzyme-needed air.

Temperature changes cause a similar change in your rate of respiration (the process of breathing). A hot shower saps your vitality and slows down your breathing. Your enzymes become similarly sluggish. A cold shower may give you a feeling of invigoration but it also involuntarily causes you to breathe at a rapid rate so your enzymes become supercharged and over-active. They work at such a top speed that were you to sit down and eat heavily after severe cold bathing or exertion, you would run the risk of indigestion.

Germ-killing enzymes.

Considering the fact that you take about 25,000 breaths a day, each pintful of air passing through your nose contains germs, sand, dust, soot, poisonous chemicals, etc. How can enzymes nullify or kill many of these infection-causing substances?

The answer lies in the moisture or mucous that is secreted by your nostrils. Coarse hair lines your nasal passages, trapping all (or most) dust and dirt. The finer particles of dust and bacteria are caught in the secretions of the nose's mucous membranes and carried along to the throat. This job is carried out by millions of tiny hairs known as *cilia*—organisms that wave back and forth in perpetual motion (like a field of wheat in a strong breeze), about 250 times a minute, sweeping all kinds of impurities before them.

As air is taken in through the nose, it is moistened by a secretion—this is a fluid (about one quart is made each day) that acts as a germ-killer. *Lysozyme* is this powerful enzyme which has the job of filtering, humidifying and warming all the air you breathe. If you breathe through your mouth, you deny yourself the protection of the enzyme, lysozyme.

Further, lysozyme, according to Herbert S. Benjamin, M.D., author of *A Little Sickness,* "is completely replenished every 15 minutes by mucous-secreting glands, and thin and clear as the wall of a water bubble, this vital impurity-disposal mucous film travels constantly toward your throat where it is constantly swallowed without your feeling it.

"Hurrying at the rate of an inch a minute, this frail mucous film, *rich in germ-killing enzymes,* is actually whipped along by millions of lashing cilia, invisible millionths-of-an-inch-thick wavy hairs which clothe the nose's insides like a velvet drape."

Dr. Benjamin cheers the power of this enzyme as a remarkable bulwark against infectious diseases. Lysozyme as well as other enzymes in the glandular secretions moistening the nasal passages, are in dire need of fresh and copious air for strength to keep you healthy.

Good ventilation is needed.

Proper ventilation is both an inside and outside job. You may think you are breathing properly but still feel ill and mentally and/or physically below par. Take an "oxygen cocktail" every morning, to fill your lungs so you will stimulate your enzymes and thereby vitalize your entire body.

Such negative emotions as worry, fright, anger, will cause a depression of bodily functions including respiration. Ever notice how you become a short-breath person when losing your temper or in a nervous situation? Your enzymes become deprived of needed air, creating a possible health hazard. Tranquility and optimism are vital for enzyme strength and function.

As for improperly ventilated working areas, persons confined indoors should step outside, or stand before an open window, several times a day and take deep breaths of fresh air. This exercise expands the dormant cells of the lungs and increases the respiratory capacity to benefit your enzymes.

Shopper's headache.

For weeks, local papers carried advertisements about an inventory sale in the largest downtown department store. It was sure to be mobbed but Lillian Fischer was going to brave it. "Prices are less than half," she bubbled enthusiastically, "so I'm saving up my money to go on a shopping spree."

Her neighbor, Adele Howard, was just as excited. "I'll get up early enough to have a nice breakfast; then I'll start out early for the store. Jim said he'll take the children on a camping trip the night before so I won't spend time in making breakfasts for anyone other than myself."

Lillian furrowed her brow. "Isn't Jim taking the car?"

"Yes," answered Adele. "What does that have to do with it?"

"So how will you get to the store? You have to be there bright and early or else the best marked-down items will be gone. You read the slogan in the newspapers, didn't you? About first come, first served. Low quantities and all that. Are you going to try and get a cab at that early hour of the morning?"

Adele was calm as she explained, "Lillian, it's only about 9 or 10 blocks. I'll walk."

"Go ahead and be a Nature girl if you want. As for me," Lillian thumbed in her own direction, "I'm not wasting time on breakst or walking. I'll drive." She offered to take Adele with her but the latter explained she would be sleeping a bit later and preferred the walk since early morning was the most favorable for fresh air.

The inventory sale was a huge success. Apart from a few expected calamities including broken glass and bruised shins, it went smoothly. Only one incident caused alarm. A young woman, by the name of Lillian Fischer, who later said she had been on line for a half-hour before the store opened, became dizzy, lost her footing and collapsed—after she had been in the store for less than an hour.

"I don't know why I felt that way," she wailed in the doctor's office. "I felt good when I woke up. Just coffee and toast for breakfast. Then I drove to the store. It was easy to wait on line, but I became sort of tired once I was inside. Then everything turned hazy."

The doctor made a simple diagnosis. "Your blood sugar level was low, Mrs. Fischer. You denied yourself a good breakfast, for one thing. Then you hopped into your car and drove straight to the store. You hardly had a chance to take deep, fresh breaths of air to replenish your sleepy oxygen cells. In addition to a low blood sugar rate, you had a 'starved oxygen' condition. Combined, they made you feel weak and you fainted."

When she told this to Adele Howard, the neighbor who took time out to fortify herself with proper nourishment, there was no undue alarm. "Well, I told you it's nice to have a long walk in the morning. I choose the side streets where there are few cars and lots of trees and lawns. I breathe deep, hold my breath, then exhale. It's like a tonic—a health tonic." She indicated her vast purchases. "See? I bought everything because I had a clear and alert mind. Fresh air is food for your mind!"

The next sale held saw Mrs. Lillian Fischer walking to the store, her face looking fresh, her eyes alert. She had a solid breakfast—and a half hour of walking in the clean, morning air had so nourished her enzymes, they rewarded her with vigor and energy.

"Shopper's headache," is the term used by J. DeWitt Fox, M.D., in *The Doctor Prescribes*. It is a condition of starved enzymes—starved for a good breakfast, starved for fresh air. It begins with a dull ache through your temples . . . the advance warning sign of enzymes that need precious oxygen.

"The fact that you were so busy you didn't even notice there were water fountains in the department store means you went dry while shopping. When we fail to drink sufficient water, a so-called dehydration headache can hit us. Women are especially poor water drinkers. To appreciate this fact, just look at the dry, parched lips of so many folks downtown. The busier we are, the more important water intake becomes. Sadly enough, it's when we are busy that we neglect to drink water."

Tension irritates enzymes.

Nervous tension causes irregular breathing that irritates your enzymes. A tight schedule means you have tight blood vessels and headaches. Enzymes react by constricting when you are tense, air and water-starved.

Here is a 3-way plan that is simple to follow in coping with tensions:

1. *Start out early.* Arise early enough to enjoy an enzyme-rich breakfast. Drink a freshly squeezed fruit juice, a glass of milk, soft-boiled eggs, 100% whole wheat bread and a fresh, raw fruit. The milk and eggs have protein that is slow in digesting so your enzymes enjoy this "delayed action" feeding. This type of solid breakfast will also keep your blood sugar at a healthy level.

2. *Drink lots of water.* Prior to your leaving home, drink several glasses of water. This helps to hydrate your enzymatic system and helps ward off headaches. Furthermore, when downtown, look around for water fountains. "Don't just take a sip to wet your lips," cautions the doctor. Drink big quaffs of water. This means 12 to 14 average swallows—not one or two. Tea, coffee or cola drinks are not suggested since they do not add water to your enzymes but contain stimulants which cause your kidneys to excrete abnormally large amounts of water. You defeat your purpose when you drink them in place of water or fruit juices.

3. *Take deep breaths.* Air pollution is an unhappy condition. City air is filled with soot, smoke, dust. In fact, standing at any city street corner will cause involuntary intake of carbon monoxide that may cause dizziness. Enzymes choke and die from poisonous air. What to do? Find a little park for a spot of fresh air while downtown, if possible, to take ten deep 'belly' breaths to wash out your lungs. Flush out the gaseous wastes, such as carbon dioxide and carbon monoxide fumes, smoke and soot. Exchange these for all the fresh oxygen you can breathe in.

Your morning oxygen cocktail.

Just as you are attentive to the food needs of your enzymes (by means of much raw fruits and vegetables eaten under proper conditions), you should be concerned with their oxygen needs. When you feed pure oxygen to your enzymes, they can electrify your entire system. Not only must you breathe—*but breathe deeply!*

A casual whiff is as useful as a nibble of food. It sustains you but little else. Each enzyme needs its maximum share of oxygen and this is possible only if you take time to *breathe deeply and properly.*

You say you're too busy? Below is a simple 5 minute breathing cocktail session that is easy, fun-filled and healthy. No one can honestly say that he cannot "be bothered" by taking this oxygen cocktail.

(A prominent health authority says that without oxygen, the starved enzymes could cause death "in the short order of three seconds!" Similarly, without sufficient oxygen, your enzymes are only half-alive—and so are you!)

Here is a special morning oxygen cocktail which is a powerful stimulant to your health-building enzymes:

1. In the open (fresh) air, or with windows open, stand with hands on your lower ribs.
2. Breathe in slowly through the nostrils, and be sure your *lower*

ribs are being pushed out. Many people breathe without using the lower rib muscles to their fullest capacity. It takes practice.

3. At the end of breathing *in* (inspiration), after taking in all the air it seems possible, take in *another whiff*. Then take one more. As you take in the last whiff, hook your fingers under your ribs and given them a tug outward.

4. Now with mouth open, let all the air out, and at the end of expiration push the lower ribs *in*. Grunt to get the last bit of air out. Do the above exercise three to five times—three times a day.

Could anything be simpler? You can vary this little exercise by breathing in rapidly and expelling the air slowly; or breathe in very slowly and exhale rapidly. It was found that singers and speakers can develop breath control with this exercise. To add still more variety, raise your arms slowly up over your head while breathing in. This expands your rib cage.

Test your own enzyme air health.

You may think your enzymes are properly oxygenated, yet they are not performing their best. Here is how you can test your own enzyme air health and find out if you are breathing properly.

Put a colored picture on the wall. Assume a good standing posture before it. Empty your lungs as much as possible. Now, breathe deeply through your nose until your lungs are filled as much as possible. Pucker your lips in a whistling motion. Slowly—very slowly—blow out all the air, thereby emptying your lungs.

Do this same test about four or five times. At the end, look at the picture on the wall. Are the colors blurred? Are the lines indistinct? This means you are improperly breathing. You need to practice deep breathing more regularly.

Before tackling any job, take deep whiffs of fresh air; you'll be able to memorize a speech more easily; you'll be able to add up a column of figures with surprising alertness. You will be better able to argue against your teen-ager's insistent demands for an increased allowance or more car privileges. Your enzymes reward you with a fresher perspective when you feed them lots of delicious oxygen.

Case of the choked enzymes.

A railroad engineer was once treated by the famed Dr. William Mayo. The engineer complained of periodical blackouts while driving his engine, and of fainting on numerous occasions. The engineer-patient was nattily attired and wore a high, stiff collar.

Dr. Mayo ventured, "Tell me, do you always wear that type of collar when driving your engine?"

"Indeed, yes. They call me a 'dandy' but I still wear it. Mostly habit, I guess."

"Are there several sharp curves on the line you operate?" asked the famous doctor.

The engineer-patient said there were.

The great physician replied, "If I pressed my finger against the side of your neck, I would interfere with the supply of blood to your brain. In a few minutes, you would faint. When you lean out of your cab as the engine goes around the curve, you are pressing your neck against the high stiff collar—cutting off the supply of blood to your brain. I suggest you get rid of those constricting collars."

The engineer-patient took the advice; the high, tight collars were discarded. While he lost some of his dandified appearance, he also lost his blackouts and faintings!

The first thing we do when a person faints is to loosen the collar or other tight garments. This gives precious oxygen to the enzymes, although few people are aware of it. No doubt, many people who feel dizzy, prematurely fatigued, or who fall asleep at the wheel and can never explain why, after a fatal accident, are choking their enzymes. You may be breathing deeply of fresh air, but wearing excruciatingly tight garments means that the oxygen supply is blocked off and cannot get past the binds.

Wear comfortable fitting garments. Women should particularly avoid the "choked enzyme" condition caused by tight girdles. These garments hamper the action of the arm and chest muscles; so do tight elastics around the middles.

Men are equally guilty by wearing tight belts; some wear winter clothing that is heavy and may offer protection from the cold, but prevents oxygen from reaching all body parts because of tightness.

Tight shoes, too, that are too small or ill-fitting cause incalculable misery and pain. You tense up. Your nerves are taut. This is hardly a favorable environment for enzymes—which, as explained earlier, react with extreme sensitivity to all internal and external circumstances.

Even though you have comfortable shoes, it is amazing how much tension relief is possible by just loosening the laces or getting out of them entirely.

Clothes should fit with comfort, not with strait-jacket tightness.

Tight belt caused constipation.

Since enzymes figure into nearly all vital bodily processes, it is reasonable to understand how constipation can be caused by starved enzymes. I remember a man who suffered from constipation as far back as he could remember. This man was a laxative addict. All concoctions, tablets, syrups, powders, etc., were to be seen in his medicine cabinet. Nightly, he took enemas.

He was examined by specialists all over the country. X-rays showed nothing abnormal that could be blamed for constipation. This man said that he considered himself lucky to experience a bowel movement just once a week. He was unpleasant to be with, for obvious reasons.

One afternoon, a visiting granddaughter chanced to remark. "Grandpa, why do you wear your belt so tight?" The little child looked at him with youthful innocence.

Suddenly, the man deduced that he ought to loosen his belt! He had fallen into a "tight belt habit" ever since young adolescence during some style fad or something. He went further and discarded his belt and wore suspenders.

Now he discovered a change. He answered the call of Nature every five days instead of seven. Gradually, it increased to every other day, then just about daily. He tossed out his laxatives and has never again needed to resort to artificial means. A tight belt had choked the oxygen (and food) supply to the enzymes and some of their functions were impaired—notably, the function to prepare wastes from ingested foods.

Relieve tension with this exercise.

Tension instinctively makes you hold your breath. You are deprived of vital oxygen for your enzymes and health interference follows. Here is an exercise you do for 10 minutes at a time, three times daily. Gradually, you work up to 15 minutes, still thrice daily. This results in easing your tension and helps you oxygenate your enzymatic system:

1. Breathe through your nose, expanding your rib cage until your lungs are filled with air—but with no strain.
2. Hold your breath for just three seconds.

3. Exhale *slowly* through your mouth on a sustained note of "OO," while your lips form a rounded "O." While it is best to make this humming sound of "OO" on a single note, you may use the entire scale or a melody to avoid monotony.

This simple tension-melting exercise will then enable you to benefit from deep, relaxed breathing. Done preferably after a meal, as it can greatly aid digestion and bowel movements.

Abdominal breathing. When you have relieved tension and want to send a healthful meal of nourishing air to your enzymes, try abdominal breathing. Letting your abdomen go in and out with each breath. It can be superior to chest breathing alone. You should not raise your chest *forcibly* while breathing. Instead, let your lower chest case and your abdomen stretch out freely for deep, relaxed respiration. The lungs expand *down* as well as out; during best breathing, most lung expansion is downward.

CHAPTER SUMMARY.

Remember, your enzymes must have a plentiful supply of good, fresh air sent to them by proper breathing. Here is a summary of the highlights of this chapter:

1. Breathe through your nose so that all oxygen can benefit by being filtered by lysozyme and other enzymatic substances within your nasal channel.

2. Sit straight, stand straight, walk straight. Square your shoulders so that your chest will expand to permit more oxygen during breathing.

3. Maintain an even temperature. Avoid extremes of hot to cold or vice-versa. Oxidation is interfered with and enzymes are thrown into havoc and a panic by such sudden shocks.

4. Walk instead of riding, when ever possible. Select streets or roads that have fresh air. If you're indoors much of the day, take a "break" for just five minutes and step outside and breathe in fresh air.

5. Eat a nourishing protein-filled breakfast after which you should take a ten or fifteen minute walk, each morning. This oxygenates your enzymatic system and gives you a wonderful foundation for the day ahead. Drink lots of fresh water throughout the day.

6. Every morning, just before breakfast, do the "morning oxygen cocktail."

7. Clothing should fit comfortably but not be as tight as a strait jacket —this means foundation garments for the ladies must be snug yet not binding; belts and shoes should be well-fitting, yet not tight so that enzymes are deprived of blocked oxygen.

8. Fight tension with humming "OO" out-breathing exercise. Do this exercise after a meal if you are excessively tense.

9. For extra enzyme health, give them "abdominal breathing" throughout the day—even while you work. Just 5 minutes—as you work at your desk, factory job or kitchen sink— means a world of difference to breath-hungry enzymes.

10. *Prevent* tension by breathing *out* slowly before you tackle any job you anticipate will be monotonous or disturbing; breathe *out* if you are jittery or have stage fright. Breathe *out* before you climb a long flight of stairs. Breathe *in* for every two steps, then breathe *out* for the next two steps. This alternating process maintains a healthy enzymatic oxygen balance.

9

HOW RAW FOODS
KEEP YOU YOUNG AND HEALTHY

Cooking destroys enzymes.

Those millions of miracle enzyme workers in your body were placed there by Nature to keep you healthy even in advanced age when your appearance should fool the calendar. Nature's children, as enzymes are frequently called, must live according to Nature. Just as they need fresh air and fresh water, they also need fresh foods. By fresh it is meant *raw* foods.

The process of cooking will destroy these enzymes so that you may be eating a wide variety of foods, yet be enzyme-short. No animal willingly eats cooked food. Man is the only creature who desires cooked edibles. Just a few moments of boiling temperature will kill every single enzyme in the food.

Of course, you are not expected to eat uncooked foods such as meats, fish, eggs, etc. Not only are they extremely unpalatable, but bacterial parasites exist in such foods and may cause internal infection if ingested. Cooking will destroy these harmful parasites. So you have to eat some cooked foods—but not *all* of your foods must be prepared by fire.

Benefits of raw foods.

The famous John H. Kellogg, M.D., in *The New Dietetics*, said "Another advantage of the raw food diet which may, perhaps, be one of its chief merits is the fact that it supplies to the colon a considerable amount of raw starch which, being digested and converted into dextrin and sugar by the unused ferments always present in the feces, furnishes the kind of nutriment necessary to encourage the growth in the colon of acid-forming or fermentative bacteria, thus combating putrefaction and encouraging normal bowel action. . . .

"The art of cookery has been used not only to render food more digestible, but more often to lessen its digestibility and to transform the simple, wholesome products of Nature into noxious, disease-producing mixtures."

Raw foods are also more effective than cooked by means of their "cleansing powers" of the digestive tract. The well-known Dr. Henry Sherman of Columbia University says in *Food and Health:*

"Fruits and many vegetables also have an important relation to the maintenance of good conditions in the intestine. This is largely because of the bulk which they impart to the residues, thus giving the muscular mechanism of the intestine a chance to be effective in keeping the residual mass moving and ensuring its elimination without undue delay, the fiber of these foods also serving to give the digestive apparatus *its daily scrubbing.*"

Anti-cooking attitudes.

A century ago, someone quipped, "God made man and the Devil made cooks!" We may well presume that this unfortunate suffered from a diet that excluded some raw foods and he blamed his indigestion upon the cooking methods, rather than on his ignorance about uncooked foods.

Another wag declared, "Just try to imagine what a powerful lever the Devil possessed when he invented cooking and persuaded the primitive savages to seek after extraneous foodstuffs which could only be eaten if they were softened and made tasty by means of heat."

Yet, superstition was rampant. In publications of 1833, it was declared that no person, whether gentleman, farmer or tradesman, woman or child, should eat any vegetable matter that had not been softened or changed by cooking processes.

It was believed that demons and evil spirits existed in raw fruits and vegetables and that cooking would destroy them; if one ate a raw item, he was possessed by the Devil.

The well-known Sylvester Graham (after whom the Graham cracker is named) took the other side of the view. He said that man's digestive apparatus would be best sustained by vegetables that were neither changed nor altered by cooking. A century and a half ago, Graham advanced the theory that cooking reduces the value of foods, and also reduces the digestive powers of those who ate cooked foods.

Since few people could eat raw meats, fish, eggs, in those days (as in our time), they subsisted on a pure vegetable and fruit diet as well as nuts, seeds, whole grain edibles, etc.—but no flesh foods. The wisest course would have been to maintain a satisfactory balance.

Graham once wrote, "If man were to subsist wholly on alimentary substances in their natural state, or without any artificial preparation by cooking, he would be obliged to use his teeth freely, and by so doing not only preserve his teeth from decay, but at the same time and by the same means he would thoroughly mix his food with the solvent fluid of his mouth.

"Again if man were to subsist wholly on uncooked food, he would never suffer from the improper temperature of his ailment. If man were to subsist entirely on food in a natural state, he would never suffer from concentrated ailment."

This promised a lot but, unfortunately, we know today that while raw foods are treasure houses of enzymes as well as vitamins, minerals and other nutrients—one must have meats and cooked foods for precious proteins and amino acids.

Yet, most people today are going in the extreme opposite—they eat cooked foods almost exclusively and rarely touch raw foods. These are the enzyme-starved folks who complain of many ailments, all the while boasting, "I eat a balanced diet." Balanced? Does the diet include raw fruits and vegetables—every single day?

Chewing benefits of raw foods.

A raw food such as celery, carrots, apples, radishes, lettuce, green peppers, etc., means that you have to chew, supplying your teeth much needed exercise.

Raw foods call for more chewing and proper insalivation and secretion of mouth and digestive enzymes. Raw foods help preserve the teeth and stomach from abuse by hot foods.

Raw foods are rich in vitamins, minerals, enzymes, etc., in the un-impaired state in which nature produced them.

Proper chewing and tasting tends to prevent overeating.

Raw foods are less subjected to canning, pickling, fermenting, tampering, as are packaged foods that are preserved, pre-cooked or otherwise "pre-digested."

Raw foods better assimilated.

Some raw foods are better assimilated than if cooked. Cooked beets, for instance, passes through the digestive system without even being absorbed properly. It has been noticed that after eating cooked beets, the calcium and iron-rich red color shows up in the waste matter. You will find that eating raw beets will eliminate much loss of this red; or, if it shows, it is in a reduced amount. The health potency of raw beets is ruined by cooking.

Child lives on raw food diet.

Several decades ago, a famine ravaged the bleak wastes of Labrador. A well-known British physician, while making a trip, came to this land and was told of a home where a 3-year old child had survived, while his parents perished.

The doctor went to the house and found that the child's parents had died of starvation—but the youngster was alive and comparatively healthy. From questioning of neighbors, the doctor learned that the family, until the end, kept some chickens with them in the house, to furnish eggs.

The mother would peel potatoes, throw these peelings to the chickens for food. The 3-year old child crawled on the floor among the chickens, ate the raw peelings. His parents, on the other hand, ate only the inside portion of these cooked potatoes—and they died. The child, undoubtedly was kept alive by the precious minerals in the raw skins of the potatoes. The skins contained enzymes, as well, and it is believed that the child was spared death because of this particular raw food.

Another interesting raw food cure concerns that of an older country woman who was beset with eczema and skin eruptions. She was told to eat as many raw foods as possible, especially potatoes, and in a short time, her condition cleared up. No doubt the high mineral content of the potato skins (raw) relieved the woman's deficiency. The minerals activated the enzymes which then became alert and could properly function in building healthy skin tissues.

Arthritis diet of raw foods.

Dorothy C. Hare, M.D., writing in the British journal, *Proceedings of the Royal Society of Medicine* (Vol. 30), describes an experiment in the

Royal Free Hospital. Arthritic patients were given a diet of raw fruits and vegetables. The 12 patients treated had such disorders as muscular, osteoarthritis and rheumatoid arthritis.

Dr. Hare, a most conservative physician, states that this diet is not a replacement for medical therapy, neither is it to be advised for other cases. She simply reports on it. The patients were given *no* other treatment except this raw fruit and vegetable diet; to add variety, they were given nuts, cream, salad oil, milk and raw oatmeal. Here is a sample menu:

BREAKFAST: Apple porridge made of grated apple, soaked raw oatmeal, grated nuts, cream, fresh orange, tea with milk and cream.
MID-MORNING: Tomato puree with lemon.
DINNER: Salad of lettuce, cabbage, tomato, root vegetables; salad dressing with oil; mixed fruit salad and cream.
TEA-TIME: Dried fruits, nuts and tea with milk and cream.
SUPPER: Fruit porridge, prune, apricot or apple; salad dish with dressing.
BEDTIME: Lemon and orange juice with hot water.

After two weeks, these cooked foods were added to the diet: vegetable soup, one egg, 2 ounces of meat, 2 ounces of bread, 2 ounces of bacon, butter, cheese and milk.

Throughout the weeks of this raw food diet, the arthritic patients were given *no* salt. All dried fruits and raw oatmeal were soaked in water. Vegetables were shredded. Nuts were crushed or eaten whole. All food was prepared fresh for every meal and served rather pleasingly.

Results? Between one to four weeks of this raw food diet, 8 patients felt better. Then, 2 others were improved but had a slight relapse. The remaining 2 showed no improvement at all.

The 8 who improved had gone home. They continued to improve to a marked degree. One was a 46-year old woman who had suffered from arthritic symptoms for four years. She had pain and swelling of the knees. Her shoulders, arms, hands, knees and legs were stiff and painful. She remained in the hospital bed for ten weeks before she started on this raw food diet. Three weeks of it and she was discharged from the hospital.

At home, this woman continued on the diet and seven weeks later reported that she had no pain whatsoever. She claimed she was cured! All of these patients lost weight and there is little doubt that obesity added greatly to the arthritic distress.

Dr. Hare remarked that results were possible because of factors (enzymes) present in raw food. The doctor referred to a Zurich physician who also used similar raw food diets in treating arthritis and results were successful "because of the absorption of the unaltered solar energy of plant life. . . . Science has so far revealed nothing . . . of this occult solar energy, as something apart from vitamin and chemical constituents (of food)."

Discounting any solar effects as well as any occult influences, we may well understand that such a diet proved healthful and even curative because the raw foods were rich in necessary enzymes. Raw fruits and vegetables as well as grain foods had lots of vitamins A, B-complex and C; the protein and fat requirements were met by the cream, nuts, salad oil, meat, eggs and cheese.

Of course, such a diet was conducted under hospital supervision and upon special cases so Dr. Hare cautions anyone against looking at this method as a cure-all.

Raw foods come first.

Each and every meal should start with a raw salad. Before breakfast, eat a freshly prepared raw fruit salad of any seasonal items. Before lunch, eat a raw vegetable salad. The same applies to dinner. Since your appetite is keen when you begin a meal, it is wise to introduce raw and uncooked foods to your waiting enzymes. Chewing raw fruits or vegetables will alert your enzymes into action, preparing them for the cooked items that follow.

Do *not* drink with your meals. The water content of succulent raw fruits and vegetables will lubricate your digestive apparatus in a natural and comfortable manner. That is another reason for the raw food prior a cooked food. Liquids, as we have seen, may hinder enzymatic function —if you bolt down water while you eat! Liquids coming from a raw celery stalk or a few leaves of lettuce are in ample supply, placed there by Nature.

While you eat, large amounts of enzymatic juices are poured into your digestive system. Drinking water will dilute these enzymatic fluids, weakening the entire process of digestion.

Avoid spices.

A fictitious thirst frequently follows a meal. This is especially true if the meal has been salty, greasy or full of spices and condiments. If thirst following a meal persists, eat some raw vegetables or fruits. The person who eats succulent raw edibles and avoids condiments will overcome this drinking compulsion either during or immediately after a meal.

Condiments are anti-enzyme. That's right! Almost all harsh condiments and so-called sauces will cause a destructive action upon enzymes. Irritating and volatile oils are abrasive to the sensitive lining of the digestive tract where the enzyme issuing glands are located.

Furthermore, condiments not only irritate the digestive system and impair their functioning powers, but they blunt the true appetite and diminish your enjoyment of foods. You taste the spices, but not the foods. Some cases have been reported wherein excessive use of spices led to irreparable stomach injury as well as impairment of the liver, intestine, kidneys, blood vessels, heart, etc.

Enzyme-happy salad dressing.

Suppose you seek a *natural* dressing to put upon a fresh, raw salad, how can you make it? Here is one that is a digestive-aid because it favors enzymatic flow and also increases so-called mouth excitement which makes the meal more delicious. This salad dressing will also make raw vegetables more interesting to those who are reluctant to eat raw foods.

Drop one egg, one tablespoonful freshly squeezed lemon juice and four tablespoonfuls of peanut oil in a large bowl. Beat with an egg beater for 30 seconds. Continue to beat and add more peanut oil very slowly while adding a pinch of celery salt to taste, until one cupful of this peanut oil has been added altogether. More oil makes a thicker dressing. For variety, a teaspoonful of pure apple cider vinegar may be added and well beaten in.

If you have difficulty in obtaining pure peanut oil as well as celery salt and pure apple cider vinegar, inquire at any health food shoppe or special diet store. Nearly all stock these natural foods or should be able to tell you where to purchase them.

Put this dressing on vegetable salads to be eaten *before* the meal.

Morning enzyme booster.

You should want to start your day off right with a good, raw cereal. For almost half a century, this enzyme booster has been used at the famed Bircher-Benner Sanitarium in Switzerland. It is served for breakfast and dinner. Known as Swiss Cereal or Swiss Mueseli, it is delicious and creamy and a real powerhouse to the enzymes, right after a night's sleep.

Here is how to make it—and note, all ingredients are raw and uncooked:

Simply soak overnight one level tablespoon of whole grain oatmeal in two tablespoons of water. Next morning, add the juice of half a lemon, one tablespoon of soya milk and mix. Quickly shred one large unpeeled apple into the mixture, and stir in a tablespoon each of honey and fresh wheat germ. Serve at once. (To increase the protein content you may add a tablespoon of chopped almonds, walnuts or hulled sunflower seeds).

If you wonder where to obtain some of these ingredients, look to your health food store as the best source.

Fruit salad dressing.

That's right—here's a delicious dressing to put on a raw fruit salad, again designed to perk up interest in raw foods and stimulate enzymatic flow:

Mix together 1 teaspoon of the following: sesame oil, apple cider vinegar, papaya concentrate. Add 1 teaspoon of freshly squeezed lemon juice. (If mayonnaise is preferred, whip an egg into this mixture). For more sweetness, add a little Tupelo honey. Mix well in a blender or with an egg beater and pour over a fruit salad dish.

The ingredients in this dressing are prime enzyme sources, as you may have noted. For hard-to-find items, again—visit your local health store.

Health condiments.

To replace such harsh condiments as salt, pepper, distilled white vinegar, mustard, etc., you can use help (a sea salt made from actual sea-weed), vegetized salt made from dehydrated celery, garlic, etc., special herbal powders, vegetable extracts and a yeast-extract. You may use these as sandwich spread flavoring agents, as an addition to soups and stews, and wherever a spice is required. Use them in moderation, though. They are found in gourmet shops and also in health food stores. Many stores also have such flavoring agents as dill, sage, caraway seeds, bay-leaf, basil, marjoram, chives, mint leaves and watercress. These natural-source spices are soothing to the palate where mouth enzymes flow and also calm and tranquil to the digestive system where the greatest supply of enzymatic juices are to be found.

You need raw food.

Yes, you must have raw fruits and vegetables for enzymes as well as valuable vitamins, minerals and other nutrients. Maintain a happy balance.

Each and every meal should be preceded by a raw fruit or vegetable salad. If possible, substitute a whole meal with a raw dish menu. (Recipes are to be found at the end of this book). Reward your enzymes with raw foods and they will reward you with health and youth.

CHAPTER SUMMARY.

Now, let us just review the essential points in this chapter:

1. Cooked foods destroy enzymes so you should eat as many uncooked foods as possible. A safe rule of thumb is to eat raw—what can be eaten raw. Eat cooked—what must be eaten cooked.

2. Those who complain of poor digestive power would be wise to increase their raw food intake.

3. Before each meal, start off with a raw salad—and chew thoroughly to prepare your enzymes for their tasks ahead.

4. Avoid drinking with your meals since this dilutes the enzymatic flow, thereby weakening their digestive powers. Complain of dry mouth and throat? Eat more succulent raw vegetables before your meal; these include celery, green peppers, carrots, raw cabbage, raw beets, tomatoes (very high in water content), red radishes, watercress, lettuce.

5. Substitute harsh acting spices with vegetized spices.

6. Begin your day with the famed Swiss Mueseli, all-raw breakfast dish.

7. Use natural fruit and vegetable dressings.

10

RIGHT AND WRONG
FOOD COMBINATIONS

Double anniversary becomes double tragedy.

"Yes, sir," beamed George Wright when they reached the campus of the university from which he and his wife had graduated exactly thirty years ago, "times have changed—but not for our old alma mater . . . and not for us, either."

Isabel Wright was every bit as enthusiastic about the alumni anniversary celebration. "Just think—thirty years ago, George, we both graduated. And went straight to the altar. My, it seems so long. I'll bet everyone has changed, as much as we have, too." She smoothed her greying hair, glanced over at George. He looked different, now, as they walked among the long buffet tables spread on the campus grounds. He had developed a paunch, his handsome face looked lined and furrowed, he was anything but the star athlete of the college's gridiron. Three decades ago, he had been a young Adonis.

"We've not changed!" declared George as he prepared to greet the old team members who were hurrying toward him on the campus. "Today, we push back the calendar. Isabel—let's stuff ourselves the way we used to when we were kids." Further talk was lost as the both of them were surrounded by well-wishers and "you haven't changed a bit" cries of delight at the reunion.

It was a sumptuous feast. The alumni association provided more than fifteen tables groaning with such delights as lobster salad, veal cutlets, fruit compote, egg souffle, Southern fried chicken, butterscotch ice cream, seven layer cakes, meat balls de luxe, canapes, anchovies, caviar, chop suey, sponge cakes, caramel candies in gelatin desserts, liquors, coffee, tea, ice cream sodas made to order.

Before the day was over, George and Isabel Wright had stuffed themselves haphazardly. They decided that they *were* as young as they used to be. To prove it to themselves and their long-separated friends, they ate with as much gusto as they did when in their late teens, three decades ago. They ate with no planning or discretion. One moment, Isabel was gobbling down the heavenly clam chowder; the next moment, she was eating caviar on soda crackers, then she started on cherry ice cream—only to start eating a shrimp salad when she had barely finished the ice cream.

George did about the same, laughing at the way he, with his buddies, would eat "backwards" as a lark, in the good old days. "Remember how we would start with dessert and finish with appetizer?" This brought a round of laughter with dares to do it again.

Toward evening, when the alumni association called everyone into the main auditorium for a speechmaking session, George clutched his right side. He grimaced, staggered and would have fallen if the hysterical Isabel did not scream out of fright. Someone grabbed the pale, perspiring George who gasped, "I . . . I don't feel so good."

Isabel felt herself growing hot and cold. She put her own hand to her forehead and sighed, "I'm rather ill, myself."

Neither George or Isabel Wright could attend the after-banquet speech affair. Both were rushed to the nearby hospital. They were not alone in the emergency ward. Others had suffered from what was diagnosed as "acute indigestion." Several later developed food poisoning . . . although the food was not found to be toxic, according to hospital authorities who investigated.

George remained confined for four days as doctors treated him for intestinal obstruction. Isabel suffered from severe indigestion which may have led to uremia (blood infection). It was a near tragedy. Had they been stricken on the speedway in the middle of nowhere, miles from a phone, and not brought to the emergency ward of the hospital, they would not have survived.

"It must have been something I ate," wondered George. The doctor who took his case history then told him,

"Not only did you eat too much—you ate wrong combinations. Ever hear about oil and water not mixing? The same applies to food."

Isabel, well enough to sit up, walked weakly to the bedside chair and sank down, exhausted. "But, Doctor, we used to eat like that when we were young."

The doctor frowned. "That was more than thirty years ago, Mrs. Wright. Nature has changed your digestive system throughout the years. I'll prepare a chart for you, showing what foods can be eaten in what combinations. If you follow the rules, you should not get sick again. Fortunately, you were rescued in time. For the sake of your health—and lives—don't go against Nature's law."

Food combining.

It is an erroneous belief that what may be eaten separately may be eaten together. Each digestive enzyme acts upon a specific food item. Introducing contrasting foods means that the digestive tract issues contrasting enzymes which work against each other. The reaction is the same as if you were to prepare a broiled steak smothered in onions, seasoned with sharp spices—and then eat it with sugar-sweetened ice cream! The protein enzymes and the sugar-starch enzymes gush forth in the stomach to digest both of these foods: the enzymes are at opposite ends. One is harsh and powerful for strong fibrous meats. The other is mild for a dairy product. Both blend together and become diluted. Result? Indigestion.

Acid vs. starch.

A prominent health authority is against the habit of eating strongly acid foods and starches at the same meal. True, the strong digestion often gets away with this practice without suffering local disturbances in the stomach, but it is often this owner of the strong digestion who suffers from the most incurable stomach symptoms in his later years.

Acid grapefruit or orange and starchy cereal or white bread toast are often eaten together. From what has been said above about starch digestion by the saliva in an alkaline medium, it readily can be understood that starches eaten with acid fruits cannot well be taken care of by either secretions of the mouth or of the stomach.

The stomach enzymes can never digest starch under any circumstances and saliva is never able to digest starch in the presence of acid. Starches *are* good food. Undigested starches, however, are *not* good food. When they remain too long in the digestive tract, undigested, they are akin to poisons.

Chew your starches.

Starches entering the stomach with other foods which prevent their digestion cannot leave the stomach until the other foods which prevent their digestion are sufficiently broken down by the stomach fluids to be able to leave the stomach.

This generally means from four to six hours, even eight hours in some cases. In such cases, fermentation of the starch is very apt to occur, resulting in the formation of carbon dioxide, alcohol and acids.

Chew starches and thereby give the mouth enzymes a chance to digest them; when properly chewed, you can take starches with an acid meal. This means you can have corn flakes with fruit for breakfast—but do NOT bolt down the corn flakes (starch food) and then gulp the fruit (acid) for your stomach enzymes are not prepared for both of them at the same time.

Speaking generally, we see that it is not *what* you eat that builds health, strength and vital youth, but what you digest and assimilate that truly matters. A healthy digestive system and strong enzymatic flow is possible when you guard yourself against improper food combinations and mental and physical conditions which disturb and impair the digestive enzymatic flow.

Each enzyme is individual.

As we have seen in previous chapters, each enzyme works upon a specific food. Compare it to a lock and a key. Each lock has a special key. If the key does not fit the lock, no reaction is possible. Worse, the key may jam in the lock and everything is ruined.

The same situation occurs when the wrong enzyme meets the wrong food. It is reasonable to understand that the admixture of different types of carbohydrates and fats and proteins in the same meal are distinctly inharmonious to the various enzymes.

There is more water or less water, a higher degree of alkalinity or acidity, a different degree of concentration of the enzyme or a total absence of enzyme, as required by the foods that you eat and in their combination.

The character of the enzyme juice corresponds with the requirements of the food to be acted upon. Carbohydrate foods need a juice that is rich in carbohydrate-splitting enzymes. Protein foods need a juice that is rich in protein-splitting enzymes, and so forth. So different in character are the specific secretions poured out on each different kind of food, that the famed Prof. Pavlov speaks of "milk juice," "bread juice," "meat juice," "fruit juice," when referring to enzymes. Here is a set of rules on proper food combining for those who want to reap the maximum benefit of enzyme power. Toward the end of this chapter you will find a handy chart describing the various foods and whether they are predominantly acid, starch, protein, etc.

Acid-Starch Combinations

1. *Eat acid and starch foods at separate meals.* The mouth enzyme, ptyalin, is weakened in its starch-digesting powers when a fruit acid food is introduced at the same time. This means that your starch food is swallowed while improperly digested by the mouth enzyme. Consequently, starch foods may ferment in your digestive tract. If you chew your starch foods properly, they receive benefit of the mouth enzyme, ptyalin, and may later be acted upon by the pancreatic and intestinal enzymes. But you must first chew all starches. Else, there is the possibility of discomfort and upset stomach. Remember—ptyalin must first initiate the digestion of starch foods, so *chew, chew, chew*—or eat starch foods at separate meals.

Protein-Starch Combinations

2. *Eat a concentrated protein and a concentrated starch food at separate meals.* The action of a starch enzyme (mouth ptyalin, intestinal amylopsin, etc.) is wholly different from the action of a protein enzyme (stomach pepsin, intestinal erepsin, etc.). When protein and starch foods are combined together, there is interference with each other. It has been noted that improperly digested starch will absorb the enzyme, pepsin, and prevent it from working on the proteins and liberating hydrochloric acid. Starch will inhibit protein enzymatic work. Digestion is both retarded, delayed and even prevented! This brings us back to the value of raw foods. Suppose you eat a meal of grilled steak and baked beans of any sort. The steak needs a protein-splitting enzyme. The cooked beans (whether lima, fava, etc.) need a starch-splitting enzyme. You have to chew your steak and this alerts your protein enzymes. But . . . cooked

beans can be gobbled down without chewing and this bolting is anti-enzyme! If you eat a raw starch vegetable and chew, chew, chew, there is less possibility of incompatibility of enzymes!

Bread or cooked potatoes are starch foods; when eaten with protein foods, there is the tendency to bolt them down, while you have to chew your protein food. Starch and protein foods require different, even opposite digestive needs. The precise adjustment of enzymatic juices to requirements becomes impossible. What to do? If you just cannot avoid eating a protein and starch food at the same meal, note this carefully: digestive enzymes work on starch. Stomach enzymes work on protein. Therefore, for better enzymatic power, the protein food should come first and the starch food second.

Eat your protein first so it can digest in the lower end of your stomach; eat your starch last so it can digest in the upper end of your stomach—while the protein food is at the bottom end. This creates a form of harmony that should aid faulty digestion.

Protein-Protein Combination

3. *Eat only one main protein food at one meal.* There is protein in most everything you eat but many foods have such tiny amounts, you can disregard them in combination. A so-called concentrated protein food is one that is basically or largely all-protein. Two different types of protein of different composition call for different modifications of the digestive secretions and different timing of the enzymes in order to digest them efficiently.

For example, the greatest protein-splitting enzyme power (rennin or lactase) is poured out upon milk during the last hour of digestion. But the strongest protein-splitting enzyme power (hydrochloric acid, pepsin, trypsin, etc.) is poured out upon meats in the first hour. Eggs receive a strong enzyme flow at a different time than both milk or meats. Faulty combining creates enzyme havoc. It is worthy of note that the health laws of Moses, presented to the Israelites, are precise and detailed on improper food combining. Described fully in Leviticus, these laws prohibit the eating of flesh and milk at the same time. Apparently, Moses was aware of enzymes—since his followers survived the ravages of plague, disease and disaster, by following these simple dietary rules.

There is the matter of timing. Eggs require different enzymatic time than milk. Meat requires more enzymatic time. Combining eggs and milk, meat and eggs, cheese and meat, milk and meat, etc., calls for secretions of different enzymes—and their dilution. Weak stomachs may benefit by eating proteins at separate meals—or wait a decent interval (about four hours) before going on to your next protein.

Protein-Fat Combination

4. *Eat proteins and fats at separate meals.* Fat has been seen to exert a distinct inhibiting influence on enzymatic secretions. The presence of oil in the stomach delays the secretion of enzymes that must act on protein foods. Fats mixed with foods delays the development of mouth enzymes, as well. Fats further lower the production of pepsin and hydrochloric acid and this inhibiting effect may last for three or more hours.

The person who eats a fat-heavy meal (usually meats) is a familiar and tragic case. He eats with gusto, washes down the heavy food with liquids, then says he is tired and wants to lie down. About four hours later, he is found in his bed . . . dead. What happened? The fat-heavy meal remained lodged in the digestive tract; protein enzymes were so weak and low (hindered by too much fat) the food could not be properly assimilated. The man's heart pounded at a furious pace to speed up sluggish digestive work and this was the straw that broke the proverbial camel's back. The man's heart was already overworked by so much eating (the circulatory system works at top speed to facilitate digestion) and to call for more heart labor to increase enzymatic flow was just too much.

Had this man eaten fat and protein at separate or decent intervals, he might have remained alive. Foods such as cream, butter, gravies, spaghetti, etc., should not be taken at the same meal with meats, chicken or other heavy protein foods. It is worthy of note that foods which normally contain protein and fat within themselves (such as nuts, cheese or milk) require longer enzymatic-digestion time than protein foods which lack fat such as de-fatted cheeses, etc.

Protein-Acid Combinations

5. *Eat proteins and acids at different meals.* The stomach enzyme, pepsin, works to digest proteins in the digestive tract. Pepsin acts best in an acid medium. It is an error that if you take an acid food, you'll auto-

matically increase the flow of more acid. *Not so.* Acid fruits require an alkaline medium for digestion! So if you take a meat dish together with fruit, it means that your enzymatic system now pours forth both pepsin (which needs an acid medium) and invertase or erepsin (which need an alkaline medium). Note this: the action of pepsin is halted by alkali! When an acid food calls for much alkaline enzymatic secretion, it means that the protein enzyme of pepsin is temporarily halted!

So, what if you eat meat and fruit at the same time? The meat enzyme is weak and powerless; the fruit enzyme is stronger. Small wonder that you complain of so-called acid indigestion.

A healthy enzyme system secretes all the pepsin needed in an acid environment to digest strong protein foods.

Introduce acid foods such as fruits while you eat protein, then alkaline enzymes are poured forth, demoralizing the acid environment, handicapping protein digestion and resulting in putrefaction. Meat enzyme pepsin is destroyed by excess acid—so eat all proteins and all acid foods at separate meals or wait about 4 hours before combining.

Protein-Sugar Combination

6. *Eat proteins and sugars at decent intervals.* "Don't eat that cake now or you'll spoil your appetite for dinner." How many parents have thusly admonished their children? Why can't you eat a chocolate cream pie while you are eating some lamb chops? If that sounds too bizarre, you may ask, why not eat a cookie *before* dinner? That is, an hour or two before dinner?

The answer is that sugar foods have an inhibiting effect upon the secretion of enzymes. Sugars are not digested to any great extent in the mouth or stomach. Sugars are enzyme-digested in the intestine. Eat a cookie or any sugar food and as soon as it is swallowed, it makes its way to your intestine. Here it waits for an appreciable time until the lactase or invertase enzyme begins to digest it. If you eat a sugar with a protein meal, it means that the sugar food is held up in your system for a prolonged period, awaiting digestion of the protein. Even if eaten shortly before the protein food, the sugar food still must await the completion of protein or starch digestion before it can be enzyme-treated. While the sugar waits, it undergoes fermentation. "You'll get an upset stomach," is Mother's stern warning to the persistent child who eyes the cookie jar or cake dish while waiting for dinner to be served. *Fermentation is a forerunner of an upset stomach.*

When sugars are consumed alone, they undergo comparatively comfortable enzyme digestion. Otherwise, if sugars have to wait for priority, they tend to ferment speedily under conditions of warmth, moisture existing in the stomach, leading to acid fermentation.

Drink Milk Alone

7. *Take milk alone.* Milk acts as an enzymatic insulator. More correctly, it is a gastric insulator. That is, the cream content in milk (especially goat's milk which health stores sell in handy cans or powder form) inhibits the outpouring of gastric juice for some time after the meal is eaten. Milk is not digested in the stomach, but more specifically in the duodenum (that 10-inch length of the alimentary canal that follows immediately after the stomach proper). Hence, in the presence of milk, enzymatic flow is mild.

Milk contains both protein and fat (cream) which means it combines poorly with most foods. When it enters the stomach, milk tends to coagulate—forms curds. These curds form around the particles of other food in the stomach, thus insulating them against the enzymatic or gastric juices. This prevents the digestion of other foods until long after the milk curd is digested. (Moses was far ahead of his time when he prohibited his orthodox Israelites from eating meats with milk at the same time. Apparently he knew, by some strange phenomenal wisdom in these pre-scientific days, that digestion was impaired, delayed or ruined by milk-meat at the same meal. Health was a survival factor to the Israelites and they obeyed every dictum, knowing their lives would be both healthy and durable by following the rules of good nutrition. *The foundation of Moses' health laws is based upon enzymes.* He recognized (although he did not name) enzymes as having the power of life and death!

What about Natural Combinations?

Protein-Starch In One Food. Nature has few, if any, pure protein, pure starch, pure acid fruits. Virtually all foods are a combination of different elements but it is noted that *one* element stands out more, the others are minimal. But, let us take a piece of bread. A slice contains both protein and starch (as well as sugar, fat, etc.) If your enzymatic systems dislikes incompatible combinations, how can you eat this bread which is basically an adverse combination? The same applies to hundreds of other foods.

When you eat bread, your digestive system pours forth what is known as "neutral" hydrochloric acid that will attack the protein content. At the same time, starch enzymes are poured forth to attack this element of the bread. When this is accomplished, then more hydrochloric acid is poured forth to digest the bread protein.

Nature's Timetable. Nature sees to it that both processes: starch digestion and protein digestion, do not go on at the same time. Nature is careful in minutely adjusting the secretions as to character and timing, and to the varying needs of the complex food substance. That is the answer to the question of *natural* vs. *artificial* combinations. Eat bread together with a meat meal, instead of a neutral enzyme flow to start things off, a highly acid enzyme is poured out at once to attack the meat—starch digestion comes to an abrupt end! Nature's combinations as placed there by herself are infinitely safer than unnatural food combinations placed on your table by an unfamiliar cook.

Charts of Foods

Here is your handy chart, telling you which foods are largely starch, protein, acid, fat, sugar.

STARCH FOODS: All grain foods, bread, cakes, cereals, puddings, white flour foods, potatoes, yams, rice, macaroni, spaghetti, noodles, navy and kidney beans, oatmeal.

PROTEIN FOODS: All fresh foods: meat, fish, eggs, cheese, game, nuts, legumes (such as dried beans, peas, lentils, peanuts), milk, buttermilk.

ACID FOODS: Just about all fruits. Muscle meats are acid-formers, but when eaten alone without combination of acid-forming fruits, digestion is favorable.

FAT FOODS: All meat fats including lard, oleomargarine, suet, etc. Fats also include vegetable oils. Dairy foods are included.

SUGAR FOODS: All sweetening agents and foods to which sugar or a sugar substance has been added. This also includes cakes, candies, sweetened desserts.

CHAPTER SUMMARY.

For the sake of your enzyme health, be aware of proper food combining. Here is a brief run down on right and wrong food combinations as described in this chapter:

1. All starch foods, whether eaten alone or in combination with any other type of food, should be chewed thoroughly so your mouth enzymes can digest them.

2. Eat acid and starch foods at separate meals.

3. Eat a protein and a starch food at separate meals.

4. Eat only one main protein food at a meal.

5. Eat proteins and fats at separate meals.

6. Eat proteins and sugars at decent intervals.

7. Drink milk alone.

8. It is well nigh difficult, if not impossible, to prepare a meal that does not include some of the above taboo combinations; so keep this thought in mind—your meal should be predominantly either protein or carbohydrate. If it is predominantly protein, then include the *natural* carbohydrates found in vegetables, etc. This means that you give up that baked spaghetti side dish and substitute a raw celery and lettuce plate when having a meat or protein dish. When eating any combination, one food should be predominant, the other very minor or insignificant.

11

YOUR VITAL
ACID-ALKALINE BALANCE

Nature has given you two separate digestive systems—you have an *acid* process in which proteins are digested by enzymes; you have an *alkaline* process in which carbohydrates are digested. Your youth and health are weighed on this delicate internal scale. To keep your youth and health, you must avoid tipping this scale in the wrong direction.

Radio and TV are constantly sending you into a dither about acid and alkaline. Advertisements plead with you to alkalinize by taking baking soda, some milk-named concoction, seltzer, little tablets, charcoal and other cure-alls that will relieve you of distress arising out of what is known as acid fermentation and gas in the digestive tract.

You can obtain freedom from such "digestive aids" or "relieve distress of after-dinner" pills by knowing the fundamentals of this acid-alkaline balance and how to maintain a happy balance, without resorting to any advertised products.

Acid, alkaline, what is it?

All foods are basically acid or alkaline. There are few, if any foods that are absolutely acid or alkaline. But some are more so than others. Scientists today have set up the number 7 as the point where a food stops being acid and becomes alkaline. You will find that some foods are spoken of as the pH rating. (This means, potential Hydrogen—when a food has a larger amount of hydrogen it is known as an *acid;* if it has a lesser amount of hydrogen, it is called an *alkali*. This explains why you may sometimes read or hear of a food having a high pH—which means, a high acid rating.)

A figure that is under pH 7 is predominantly acid. A figure that is over pH 7 is predominantly alkaline. You can decide whether a food is acid or alkaline merely by taste. Milk, for example, is smooth and rather tasteless so it is alkaline. Lemon juice is so acid that it puckers up the inside of your mouth. Bread, however, is smooth and sweet-tasting so it is obviously quite alkaline.

Blood in a healthy person has from pH 7 to pH 7.8. Anything above is alkaline and this is a danger sign. Anything too far below is too acid and this, too, may be hazardous.

Hydrochloric acid decline.

It is noted that when folks reach the 40's, there is a slow and slight decline of hydrochloric acid in the digestive system. This creates difficulty in enzymatic digestion of protein foods. That is why a younger person can eat heavier foods and enjoy greater enzymatic power than one who is in middle-life.

Imbalance. Dr. Robert Jackson says, "Physicians speak of acidosis . . . but what they mean is that there is a relative acidosis, or that the body contains relatively more acid elements in proportion to alkaline elements than normal.

"Alkalis are soothing to cellular structures and the tissues and organs built out of cells. Acids are irritating to these structures. This is one reason why human blood must never be acid, for if the blood is acid, the cells and organs must never be acid, then they would die. Because the blood must never be acid, it must contain more alkaline elements than acid elements, for alkalis neutralize acids by changing them into harmless elements."

Symptoms of imbalance.

"If acids predominate strongly over the bases (alkalis) for considerable periods of time, the alkalinity of the blood will be gradually reduced and acidosis will result," says Dr. E. V. McCollum. *(Newer Knowledge of Nutrition.)* "The most common symptoms of this type of acidosis are lassitude, malaise, nausea, sometimes vomiting, headaches, sleeplessness, weakness, and loss of appetite.

"The muscles ache, the mouth becomes acid resulting in injury to the teeth, the stomach is stagnant and sour, the urine is strongly acid as is also the sweat. The latter may be acid enough to injure rapidly silk fabrics or to discolor jewelry worn next to the skin. Some eminent physicians now believe that the diseases of the blood vessels which are responsible for high blood pressure, kidney disease, gangrene and apoplexy are the result of prolonged injury due to eating excessive amounts of acid-forming foods. It happens that the foods most prominent in the diet of many Americans of today are strong acid formers."

What about acid fruits?

This was asked by a young woman who complained of feeling ill, despondent, nervous. She ate much fruit and also freshly squeezed raw fruit juices. "What if I restrict my fruit intake? Will this bring down my high acid rating?"

"*No!*" was the startling answer. "You will be surprised to know that while fruits such as lemons, oranges, grapefruits, pineapples, are acid when you eat them—by the process of enzymatic digestion, they are turned into alkaline substances!"

In the process of metabolism following enzyme digestion, some foods yield acid end-products, others yield alkaline end-products; some others yield neutral end-products—that is, milk becomes neither acid nor alkaline but remains neutral.

Lemon, orange and tomato juice are examples of foods that are acid when eaten, but not acid-forming when enzymes digest them; the organic acids (mainly citric acid) which are contained in fruits are converted to carbon dioxide during enzyme metabolism and eliminated by way of the lungs; what is left is called an alkaline forming element.

Keeping down acid.

Assuming you have what is known as acidosis or an upset stomach caused by too much acid. What to do? Extremely hot or cold foods tend to irritate the stomach membranes and cause an excess flow of hydrochloric acid. See that your foods are neither extreme.

Protein foods, especially meats, call for a lot of hydrochloric acid to be produced by your enzyme system. To control any excess, start a meal with a soup or a broth. The fragrant aroma will call forth acid that will remain in the stomach to digest the meat that is coming. Relax *before* and *after* meals. This relaxes your enzyme system and enables it to digest foods properly with a normal amount of acid. A nervous or tense condition calls forth a greater supply of acid.

Regulating alkaline level.

Nervous hysteria or conditions that forced an increased or over-breathing condition (running in a marathon race or any shock condition where you just can't catch your breath) may create alkalosis—too much of this substance. If you take sodium bicarbonate or any of those popular "alkalinizers," you also run the risk of tipping the balance and making yourself too much alkaline. This leads to faulty digestion and improper enzymatic flow.

Acid to alkaline.

In just 3 minutes, you can turn an acid food into an alkaline forming one. How? *Chew!* Mouth enzymes nullify the acid content of the food and weaken it so it becomes more alkaline. *Chew* all foods that can be chewed for 3 minutes. Then swallow.

This is important as you do need a balance in your system. Some enzymes work in an acid medium, others in alkaline. Pepsin which breaks down protein in your stomach needs an acidity of pH 1.8. Trypsin (the pancreatic enzyme juice needed for starch digestion, requires an alkalinity of pH 8.2.

Children fall sick.

They were the pictures of health: two youngsters, a boy and girl, aged 9 and 11, respectively. But within a month after they started in their new classes (separately), they began to grow restless. The boy said he had a tummy ache; the girl complained she felt headachy. No sooner was she brought to the office of the school nurse than she experienced nausea. This might not have caused undue concern except that in the same day, the girl's little brother was brought in, feeling weak, with no appetite.

The nurse remembered seeing them a month ago when they first enrolled in school and were given examinations. How was it possible that they should take such a turn for the worse—and both about the same time?

She questioned the anxious mother. "Tell me about their diets. What are they eating, generally speaking?"

The mother replied, "Well, for breakfast I give them a cereal, eggs and toast. Then, for lunch, a meat, egg or cheese sandwich. At home, I feed them a meat dish, noodles or spaghetti. Dessert is a favorite pudding." Further questioning revealed that she used a little oil or butter in the meal preparation. "Is there anything wrong?"

"Most certainly," affirmed the school nurse. "That diet is good except it will produce a very strong acid reaction in the body. It does not happen overnight—but it's a gradual change."

The nurse suggested stepping up a lot of fruit, nuts, vegetables and as many raw and fresh foods as possible. Within a matter of three weeks, the children had recovered and were back to their original state of health.

Enzymes need a healthy acid-alkaline balance or else they cannot digest food properly and you are denied valuable vitamins, minerals, amino acids and other trace elements that have not been extracted during the digestive process.

Avoid "acid remedies".

Acidosis is a condition which you produce yourself by poor eating habits. Therefore, improve your eating habits and relieve this condition. What about soda mints or taking a pinch of soda almost any time of day or night? Most folks over 40 fall into this vicious habit. I call it vicious because they take these remedies to relieve what is called indigestion. These remedies promise to counteract the excess acid in your stomach for a little while. But—acid is needed to metabolize or digest these remedies and this creates an even greater imbalance.

Acid-Alkaline Food Charts

Acid-forming foods:

Bread, cornmeal, eggs, fish, flour, lean meat, oatmeal, oysters, rice. Most items in the cereal and meat groups are included. All grain foods. (Butter is acid-forming, only in excess, neutral in moderate quantities.) Egg whites, cheese, nuts such as filberts, peanuts, walnuts, cranberries, plums, prunes.

Alkaline-forming foods:

All fruits, all vegetables, egg yolks, most vegetables. Examples, apples, bananas, dried beans, snap beans, beets, cantaloup, carrots, lemons, onions, oranges, potatoes, sweet potatoes, tomatoes and turnips.

Neutral foods:

Cream, cooking fats, starches, sugars. These are refined and have so little mineral matter left that they are neither predominantly acid nor alkaline-forming. Milk has much mineral content but the composition is such that enzymes turn it into a slightly alkaline substance.

Avoid Baking Soda!

Baking soda has been seen to cause a state of upset in your normal acid-alkaline balance. Baking soda does produce an immediate neutralization of the acids in the digestive tract, but this also begins a vicious circle. The soda now proceeds to provoke the stomach and your enzymes into issuing acid enzymes at an accelerated pace. This causes "acid rebound" that calls for an additional soda intake to outpace the acid march.

Baking soda keeps enzymatic gastric secretions in a heavily alkaline solution that frustrates what should be a healthy digestive process. This frustration is caused by a mildly acid fermentation of food in these enzymes. When baking soda blends with enzymatic juices, there is an automatic production of carbon dioxide gas within the stomach. This gas creates a pressure that is harmful as well as dangerous to those who suffer from stomach ulcers.

Avoid raw baking soda and any foods made with baking soda. Select foods which use *yeast* instead of baking soda. Baked goods should say "made with yeast" on the package.

Some health food stores have special baked goods counters offering cakes, breads, rolls, cookies, etc., that are free from baking soda and made with yeast. If the label does not so specify, ask the salesperson. If in doubt—don't buy!

What is exact balance?

Dr. Jackson answers, "Food scientists have discovered that to maintain the proper balance between acid-forming and alkali-forming foods, and thus assure a non-acid or an alkaline blood, a ratio of about 20% acid-formers and 80% alkali-formers should be at least approximately maintained." So, this establishes a 20 for acid and 80 for alkaline ratio.

This means that if you divide your day into five portions of equal bulk, one portion should be acid-forming and four-portions should be alkaline forming. One part and four parts.

By controlling the intake of acid-forming and alkaline-forming foods, you should be able to avoid the possibilities of an imbalance and the accompanying distress symptoms.

If you are concerned about your enzymatic powers being impaired by any imbalance of the acid-alkaline ratio, you might discuss this with your physician. By means of a blood and urine test, he can tell you whether you are too acid or too alkaline.

CHAPTER SUMMARY.

1. If you have such symptoms as sour stomach, strongly acid-smelling perspiration or urine, as well as constant headaches, nausea, insomnia and poor appetite, visit your physician for a complete check-up. Find out your acid-alkaline ratio and seek your physician's advice for improvement.

2. Eat much raw fruits, freshly squeezed juices, to obtain valuable alkaline substances.

3. Keep calm and avoid circumstances that cause increased respiratory activity. This may cause an increase of alkali.

4. Avoid baking soda or remedies for acid indigestion. Habitual usage causes a build-up of carbon dioxide gas in the stomach. These so-called remedies only cause an increase of acid flow since acid is needed to digest or assimilate them!

5. Baked goods and other foods should be made with yeast, rather than baking soda. The label states this plainly. Ask the salesperson. If you are in doubt—don't buy.

12

DISCOVER
SEEDS AND NUTS

Seeds build life and health.

Bible Food. "And God said, Behold, I have given you every herb bearing seed, which is upon the face of the earth, and every tree, in which is the fruit of a tree yielding seed; to you it shall be for meat." (Genesis 1:29.)

Here we read that at the birth of the world, seeds were known to be as vital to life as life itself. Today, speaking botanically, we know that a seed is a mature ovule, consisting of an embryonic plant together with a store of food, all surrounded by a protective coat. Entire civilizations were built and maintained because of seeds. The ancient lands of Egypt and Mesopotamia could develop because of their cereal grain crops, which were seeds. The civilizations in ancient America were the product of those Indian races who knew best how to grow corn, another seed product. Ceres was the Roman goddess of growing vegetation and seeds, and her name is associated with a seed-grain food—cereal.

Enzyme power of seeds.

A seed is much more than an ovule from which a living plant is reproduced. It is Nature's own storehouse of valuable vitamins, minerals and substances which spark their function; those substances are enzymes. Locked within the tiny seed are "activators" or enzymatic substances which act upon vitamins and minerals causing them to perform their precious functions within the body.

Without seeds, life might perish. Seeds contain nearly all essential food elements. If this were not so, seeds could not create life! The value of the enzymatic power of seeds was lauded by the noted plant authority, Luther Burbank in *Partner of Nature*. "Fruits ripen, not to make food for us, but to encase and protect the seeds inside—pips or pits or kernels. But we pay no attention to Nature's purpose and revel in the delicate flavors and delicious flesh of apples, pears, peaches, tomatoes, melons and all and throw aside carelessly the seeds that the plant went to so much trouble to build and in which it stored the life-giving germ and a reserve of starch to help it start in life again as a baby plant."

The seed accomplishes the perpetuation of the species. The seed is the vehicle for storing life's reserves. The seed is the crucible in which life's alchemy weaves its strange magic. Nature has placed within the seed a concentrated source of enzymatic power which nourishes the emerging plant. The seed's enzymatic power is able to form a root, stem and several leaves. This is just a start. *If a seed can do so much for a plant, how much more can its enzymatic power do for you?*

Seeds in history.

In Bible times, seed food was an essential part of the diet. Dill and cumin seeds were so valuable that tithes were paid with them. Whenever the Romans finished a heavy, rather gluttonous meal, they would eat spice cakes flavored with aniseed. In the Middle Ages, seed cakes were a "must" in most homes and palaces. Others believed that the eating of seeds would improve memory. It was a popular fad that one should eat hazelnuts for nine days, beginning with 6 and adding 6 more each day; or, eat pepper seeds for nine days, beginning with one seed and doubling the dose until it reached 256 seeds on the ninth day; a special "seed building health remedy" consisted of equal amounts of ground cloves, long peppers, dates, ginger, galanga-root and muscot nuts, then to be beaten with olive oil into a paste and eaten every morning before breakfast.

Today, we may chuckle at these weird seed concoctions but it illustrates how valuable these foods were regarded in the past. But we do know that the body is nourished by the food that is eaten and the power of the enzymes in that food. So seeds, with their concentrated enzyme source can be regarded as being among the most valuable foods in the diet.

Nutrients in seeds.

Plants grown from seeds are rich in essential B-complex vitamins as well as valuable Vitamin E. A deficiency of these vitamins, and a weakness of the enzymes which both depend upon and spark their function may lead to serious digestive disturbances, muscular twitching, forms of palsy (paralysis agitans) as well as numerous nervous disorders. Pyridoxine is a valuable B-complex vitamins found in seeds which unites with enzymes to build strong nerves.

Vitamin E which is needed for healthful functioning of the reproductive and birth processes is also found within the raw seed; enzymes cause this vitamin to utilize itself and strengthen the same reproductive process and also to aid in strengthening the heart.

The very heart of the seed is rich in those valuable unsaturated fatty acids—enzymes need these acids to help in the proper emulsification of cholesterol, the fat-like substance which has been seen to collect on the blood vessel walls and predispose a condition responsible for arteriosclerosis. Lecithin, found within the fatty portion of seeds, is taken up by the enzymes and used to break down cholesterol.

Seeds are rich in calcium and phosphorus, both minerals which combine with protein (this process is helped by enzymes) to then aid in assimilation and conversion into valuable amino acids. These minerals need the valuable B-complex vitamins to influence enzymes to help them into functioning. You can see how each nutrient depends upon another and how all of them need enzymes—just as enzymes need them, in turn! It is in the seed that we find nearly all of them, working in a happy health building manner.

Protein is the food element which makes up your brain, your nerves, your arteries, blood, tissues, skin, fingernails, body fluids, etc. It would be difficult to find any part of the body that is not influenced or helped by protein. Seeds are among the highest sources of protein. Nearly all amino acids are found in many seeds. There are some who maintain that you can live exclusively on the protein and amino acids in seeds. This is questionable because there are, as yet, unidentified amino acids existing in meats and fish so the seed has not yet been officially declared a perfect food. But it comes so close, that no one can expect strong or healthy enzymatic power without including seed in the diet.

Raw seeds vs. roasted seeds.

The raw seed contains the greatest store of enzymes as well as vitamins, minerals, proteins, etc. When a seed is roasted or otherwise treated to heat, it loses its highly valuable enzymatic power. Heat destroys enzymes so if you want to reap the maximum benefit of enzyme-power seeds, eat them raw.

Raw wheat seeds, for instance, are very delicious if you eat them in this manner: take a handful in your mouth. Keep them there for about 3 or 4 minutes. Ensalivation will make them thoroughly softened; then these raw wheat seeds become a delightful chew. You must thoroughly masticate these raw wheat seeds, not swallowing until they are nearly a liquid. This way, you obtain the most of their enzymatic and nutritional powers!

Wheat seed is rich in valuable enzymes needed for digestion. Cooking will kill these enzymes so eat all seeds raw!

Enzyme electric power in seeds.

"Want to see the real electric power of the seed?" asked a friend of mine when he invited me to his upstate country farm.

This was something new. "Seeds have an enzyme-electric power," I conceded, "because they spark the function of many vital body processes. But I've never seen seeds perform this electric power, except if you mean in a laboratory."

My friend promised me an interesting observation. At his farm, I saw his fields of sunflower plants, barley, rye, corn, etc. The crops were enjoying the benefits of a sun-drenched day.

"See those sunflowers?" he pointed. "The head contains those valuable sunflower seeds you're always talking about. Throughout the day, the head of the sunflower plant follows the sun, almost like a living head, drinking in the sunshine. I'm going to bring out a portable electric testing machine and show you something."

He brought the little machine into a field of sunflower plants. "Watch the dial," he indicated. "I'm going to put this 'conducting bar' close to the sunflower plant."

It was amazing. I watched as the dial began to swerve and move with a power of its own. "I'll bet you have some hidden transistors or activators inside that machine."

"Have a look," he invited.

My curiosity was satisfied. There were no batteries, transistors or other electrical apparatus within the machine. The electrical reaction was sparked by the head of the sunflower plant which had a high electric potential of its own—derived from the sunshine.

If a seed bearing plant can give off such a miraculous power from the outside, imagine what it will do when blended in with the body and becoming part of the body processes.

All seed bearing plants growing outside, exposed to the sunshine, have this electrical or enzyme power. Plants growing under the ground (such as the potato, mushroom, etc.) have a smaller enzymatic power.

Buy raw seeds.

Remember, all seeds should be purchased raw if you want to derive their maximum values. Vacuum-packed seeds are good, too. At most health food shoppes, you will find a wide variety of different seeds in a raw and toasted condition. Select raw seeds and eat them regularly. As for toasted seeds, these are very flavorable and have important potencies of vitamins and minerals, as well as proteins—and while their enzyme power is weak because of their toasted condition, you can enjoy them, too—but include raw seeds which should come first in your enzyme building plan.

How to eat seeds. Since the precious unsaturated oil content is so high in seeds, this food may turn rancid if not fresh. So buy them to be eaten within a short time. Use them to munch on between meals. Add seeds in a tossed salad. Sprinkle seeds over soup, fruit or vegetables. Add seeds to yogurt.

Grind up seeds in an electric blender and stir in a freshly squeezed fruit or vegetable drink.

Different Types Of Seeds

Aniseeds.	Place in apple sauce, stews. Used by the ancient Romans as a digestive aid. Stomach-soothing and enzyme rich.
Apple Seeds.	Rich in Vitamin E which combines with enzymes to strengthen the heart muscle and also improve the skeletal structure.
Barley.	A cereal seed, used as a hot poultice in ancient days to prevent infection from spreading. Known in China almost 21 centuries ago. The Swiss (a hardy, healthy nation) chew raw barley seeds as part of their daily diet.
Basil.	Seen to invigorate some people; it has a valuable magnesium content which soothes the entire vascular system when so inspired by enzymes.

Beans.   It is said that Columbus ate cakes made of pounded corn and bean, both of which may be eaten raw. Minerals within raw beans combine with its enzymes to help maintain body water balance. Eat raw beans wherever possible.

Berry Seeds. Strawberries, raspberries, loganberries, etc., are rich in raw seeds. When berries are eaten raw, in the morning, the enzymes stimulate a natural bowel movement and you can enjoy freedom from constipation without resorting to laxatives.

Buckwheat.   While buckwheat flour is delicious in pancakes or breads, it has to be refined and this destroys enzymes. Raw buckwheat is rich in valuable rutin which combines with enzymes to maintain an even blood pressure rate. Buckwheat should be eaten raw, by the handful, liquified by mouth enzymes, then swallowed.

Cantaloupe. Yes, you've seen these seeds in the core of the fruit and have discarded them. *Don't!* These seeds have high protein content. Enzymes take these proteins and aid in calcium absorption, in the formation of hemoglobin in your bloodstream. Cantaloupe seeds are soft and you should eat them with the fruit.

Caraway.   Remember Shakespeare's *Henry IV*, in which Squire Shallot invites Falstaff to a pippin and a dish of raw caraways? Today, the Scotch take a plateful of caraway and munch while sipping teas. Or, they dip a buttered side of bread into raw caraway seeds and eat. Sprinkle on applesauce, soups, with baked potatoes and salad dressings.

Carob.   You do not actually eat the seeds of the carob; rather, its seed pods. These carob pods are the fruit of the carob tree, related to the honey locust. It is popularly known as St. John's bread. Rich in many nutrients, it is a wonderful sugar-substitute. It has a delightful chocolate taste.

Coconut.   The biggest seed we know of. The entire coconut (except its outer brown skin) is the actual seed! Coconuts grow near the water so they have more minerals than seeds grown inlands. The richest source of enzymes are found in the brown skin that clings tightly to the white meat. Coconut is the main staple food item in many tropical countries.

Corn.   The most famous and favorite American grain. Raw corn —right from the stalk—is a wonderful enzyme and nutrient source.

Dill.   An old herb, its seeds were chewed in church by many of our ancestors. It provides a distinctive flavor when properly chewed.

Fennel.   It is believed that fennel seed helps in problems of upset stomach. The enzymes within the fennel seed are valuable for digestive powers. Eat these seeds raw.

Fig.   Look in the fig's "wrinkled" folds and you can see the tiny, crunchy specks which are the seeds. You must chew these seeds very thoroughly to release their enzymes or else they will just pass through your system without being assimilated.

Flaxseed.   A high source of unsaturated fatty acids. Seems to have a laxative effect.

Grape Seed.   Eat fresh, raw grapes, chew them very carefully so they can be digested. Good mineral source.

Lentils.   The soup is popular but cooking will destroy enzymes. One of the oldest seeds cultivated, you may mix them with tomatoes, onions, salads . . . raw, of course. An electric blender will pulverize them into a powder to be sprinkled on salads.

Millet.   Used as a cereal, today, it is a prime source of calcium. Chew raw millet seeds to obtain benefits of their enzymatic powers.

Nutmeg.   A popular flavoring ingredient. Mace is a part of the nutmeg seed which ancients used for antiseptic properties. Chew and swallow raw nutmeg.

Pepper.   Not to be confused with powdered pepper or the harsh condiment. Pepper is the berrylike seed of the pepper plant growing in a hot, humid climate. Rather volatile, the seed should be eaten sparingly.

Psyllium.   These seeds are known for causing a natural bowel movement. Enzymes within the psyllium seed cause it to swell, then create a bulk in the intestine. These enzymes cause a certain amount of lubrication, too.

Senna.   Another natural laxative, often infused in whey and then boiled.

Sesame.   We all know of sesame seeds in halvah and Turkish delights. Sesame seeds almost equal sunflower seeds in their enzyme power. They can be eaten as a between-meal snack, or ground finely and mixed with coconut.

Pumpkin.   Famed pumpkin seeds contain protein, unsaturated fatty acids as well as minerals which unite with enzymes to maintain a healthy hormonal function as well as a healthy bloodstream. Chew and eat pumpkin seeds daily.

Sunflower.   Unique with its heliotropic powers—it follows the sun, drinking in all the vitality of the sun, imparting this strange electrifying power into its seeds. Sunflower seeds have high amounts of vitamins and minerals. The protein content of sunflower seed meal is about 52.5%, which makes it more potent than concentrated meats. Sunflower seeds are a must!

Squash. The Maya Indians used the seed of the pepitoria squash as a means of sustenance. Squash seeds are high in unsaturated fatty acids as well as enzymes and should be eaten unroasted, unsalted and uncooked.

What about cooked seeds?

Yes—you should definitely eat cooked seeds, toasted seeds (these taste very crunchy and delicious) and foods made with baked or roasted seeds. You will benefit by the protein and vitamin-mineral content as well as the unsaturated fatty acids. But . . . and this is a *big* "but,"— you will not derive maximum enzyme power from cooked or otherwise heated seeds. So maintain a balance—with raw and untreated seeds forming the larger part of your balance.

Value of Nuts

What's a nut? A nonsplitting, one-seeded fruit, with a hard, woody shell. Picture a food that grows wild, is free for the picking, needs little care while it grows, is harvested by picking up from the ground; this food needs no processing, no cooking, no refrigeration, no preservatives. Does this sound like the ideal food we all think about? Well, this food has been here, waiting for us; slowly, we have come to recognize nuts as our best and most practical foods.

Nuts are high in protein, unsaturated fats, minerals, vitamins, etc. Nuts may not be complete proteins, although they are high-protein. This means that the nut does not have all necessary amino acids, but have so many of them that they are essential. This rich protein source combines with enzymes to prevent spoilage or decay. They should be eaten in an unroasted state to save their heat-sensitive values.

Can't digest nuts?

"Sorry," a friend pushed away a bowl of nuts when having dinner at my house. "I know they're delicious but nuts give me an upset stomach."

"You mean you can't digest nuts. That's because you've been eating nuts that are prepared in a way to cause trouble." I went on to explain that most nuts in the market are overroasted in deep fat and heavily salted. In this form, they are less digestible than in a natural state. Also, they are often eaten at the close of a heavy meal. Because they're so rich in proteins and enzymes, they become too heavy a burden that way.

My friend made a wry face. "But you want me to eat them right now!"

"If you want to. Our dinner is very light, mostly vegetables, no meat or concentrated protein. And these are *raw* nuts. They're as natural as Nature made them."

My friend took my dare. She ate a reluctant half-handful of nuts. Later she called to say, "You know, fresh, untreated nuts are wonderful as dessert—after a mild meal. From now on, I'm going to be a nut fan."

Minerals in nuts.

Many nuts have high iron and calcium sources. The almond and the filbert supply a larger proportion of iron than does an equal amount of beefsteak and several times the amount of calcium supplied by meats of any sort.

Nut proteins resemble those of milk so closely that they were long known as vegetable caseins. Furthermore, nut fats are more readily digestible than most animal fats and are less likely to decompose in the alimentary tract. Enzymes within nuts see to it that freshness and purity remain high.

Pound for pound, walnuts, almonds, pecans, filberts and other nuts rate well with leading flesh foods in protein content; almonds and walnuts exceed whole milk in protein-enzyme content.

Treated nuts.

The cashew nut (comes from India) is shelled. How? It is heated in liquid so the shells become brittle. English walnut shell is often loosened by exposure to harsh ethylene gas. Almonds are bleached by being dipped in lime chloride. Bleached pecans are often dyed.

Nut blanching—that is, the inner skin is removed—is done by soaking the nut in hot water. English walnuts and pecans are dunked in hot lye, then rinsed with an acid. Another enzyme-destroying process is to pass the kernels through a heated solution of sodium carbonate and glycerin; skins are removed by a gush of water, and the nuts are then dunked in a solution of citric acid.

All of these processes make nuts more edible, more convenient to eat. But they are so low in perishable nutrients, not to mention the total loss of destroyed enzymes, they will not do for you what Nature intended—to build health.

I tried a home experiment and cooked macadamia nuts in hot oil for just 13 minutes. I found that 15% of the valuable thiamin content (nerve building B-complex vitamin) had been destroyed. Imagine what other depletions are made by cooking.

What to do? Eat nuts whose food value have been preserved. How? Get nuts that are neither shelled, roasted or otherwise treated. Eat nuts by shelling them yourself. Avoid fancy, toasty-smelling nut and confectionary stores. Seek out special diet shops, gourmet shops, some bakeries, health food stores, and ask for raw, untreated nuts. Is it worth the effort? Try a few almonds, right out of the shell, and you'll never go back to the bleached or dyed almonds.

Nuts in their shells keep well. Those shelled nuts will turn rancid, particularly during warm months. Keep all nuts in their shells until ready to be eaten. That way, their rich enzyme content remains fresh and ready to build your health. Keep a bowl of unshelled nuts on the dining room table, together with a nut cracker; make it a trademark in your home.

Nature's enzyme food.

If there is one food we may look to as an "enzyme food," it is the nut. Day after day, through spring, summer and autumn, the magnificent life-giving sun drives the river mists before it and sends down through the softly whispering foliage a thousand shafts of burnished gold that drain the nectareous dew-drop from its chalice and kiss the nut until its youthful, mineral-laden sap changes to delightful food beneath their passionate caresses.

It takes months of sunshine to perfect the nut; when completed, it is a veritable storehouse of minerals, protein, unsaturated fatty acids, vitamins, and the life building enzymes. Packed in a Nature-made, waterproof and air-tight shell, the nutmeat comes to us clean, wholesome, and chock full of enzymes. Hermetically sealed, the nut does not become contaminated and spoiled as will other foods. Nuts are free from waste products, are aseptic and do not readily decay, either in the body or outside of it. Why is this so? Because enzymes protect the nut—as they will protect you against the ravages of illness and disease.

Nuts have been heralded as choice substances, capable of sustaining life. The nut is the choicest aggregation of the materials essential for the building of sound, human tissues, done up in Nature's own hermetically sealed package. Further, nut oil is regarded as the most easily digested and assimilated of all forms of fat. Enzymes in the raw and untreated nut see to it that the oil content is beneficial and health giving. Nut fat exists in a finely divided state; when you chew nuts, enzymes cause the production of a fine emulsion so that when nuts enter your stomach, they are in a form that is favorable for digestion.

Enzymes see to it that nuts are rich in unsaturated fat, ready-made, prepared and pre-digested—for circulation throughout the entire lymphatic system—Nature has done all this for you.

Nuts and Their Enzymatic Powers

Let us look at some of the best known nuts and their enzymatic powers which you can enjoy by eating them raw and untreated:

Acorn. A farinaceous nut produced by the oak tree. The original California Indians thrived for hundreds of years on an acorn diet.

Almond. A fine nut, higher in its phosphorus content than most others; has much calcium and potassium.

Brazil Nut. Rich in unsaturated fatty acids, has a valuable enzymatic power which enables this oil to be assimilated in the bloodstream.

Cashew. Not really a nut, but the seed of the cashew apple. Unlike other fruit seeds, it grows outside of the apple at its lower end. Benefits by being sun-drenched. Rich in valuable Vitamins A and D needed for a healthy skin and strong bone structure.

Chestnut. Its shell is thinner than that of most nuts; has appreciable vitamins and minerals and high enzymatic powers.

Cocoanut. The meat and milk of the raw cocoanut are brimming with enzymes. When sprouted, the "milk" of the cocoanut is transformed into a snow-white, sponge-like ball that is sweet and delicious. Cocoanut oil is used in place of butter.

Hickory Nut. You would have to remove its thick, hard shell and this makes it unpopular; yet it has very high protein and also much enzyme power. Remember when we would gather and eat the hickory nut before the fire in winter? Eat hickory nuts after shelling them and you'll enjoy the fresh, tangy taste.

Peanut. While not nuts, we know them as such. Peanuts do not grow as tree fruits but are legumes, like soy beans and peas. They grow on a peanut plant. As the plant matures, the branch tips bury themselves in the ground. Peanuts are formed underground. When the plant withers, that means you dig up the peanuts. Peanuts are delicious when taken right out of the shell. To store, keep in the shell, in a covered container at about 35° F. They keep this way up to two years with little food value loss; also, known as the ground pea, ground nut, goober, etc. Peanuts are rich in precious amino acids. Did you know that 100 peanuts give you 100 grams of protein? Enzymes are responsible for the high B-complex and vitamin content of peanut.

Pecan.   King of nuts, a native American—a low protein nut, it has a highly digestible oil content. When properly chewed, it is felt that even those with very weak stomach powers will benefit by the pecan. The enzymes in the pecan make this nut a desirable food for those who have digestive disorders. Many chronically underweight persons pick up added pounds when consuming pecans in great quantities.

Pignolia or   High in unsaturated fatty acids and enzymes, this nut is
Pine Nut.   also protein-rich, easily digestible. Enzymes create good vitamin and mineral content.

Pistachio.   Greenish in color—the greener they are, the more powerful the enzyme content. High-protein. Because it contains no indigestible cellulose, the pistachio nut is all food.

Walnuts.   Various varieties include the English walnut, the black walnut, the butternut, etc. Again, these nuts have prime enzyme power and also much unsaturated fatty acids. Some confuse the walnut with the chestnut which is another nut to be enjoyed—but eaten raw!

How to eat nuts.

All nuts, especially the almond, must be chewed to a creamy consistency if they are to be well-digested and for you to obtain the maximum enzyme benefit. The all too common failure to thoroughly chew nuts is partly responsible for their reputation of being "difficult to digest." Eat the nuts fresh from the shell. Buy nuts in the shell, of course.

Benefitting from seeds and nuts in other forms.

You must bear in mind that when seeds and nuts (and all other foods) are subjected to heat whether by roasting, boiling, dipping in hot fat, etc., the enzyme content is lost. The protein, vitamin and mineral contents are somewhat depleted, but not entirely lost. This means that you *must* eat seeds and nuts in a raw state for enzyme power. To vary your seed and nut diet, try them in other forms: *in flour* to be used for baking, *in pulverized form* to sprinkle over salads, in soups, over desserts, in yogurt, stirred in any beverage, hot or cold; try them in the form of butters—*peanut and cashew butter* are well known, or as spreads on whole wheat bread. When seeds and nuts are processed to create flour or butter, etc., the enzymatic content is lost but the other nutrients are retained to an appreciable level. Again—strike a balance. Eat seeds and nuts raw, whenever possible. Use them for cooking or as spreads or beverage flavoring, to obtain variety.

Health food stores have such products as sunflower seed flour, sesame seed flour, soybean flour, peanut flour, to name just a few. They also have butters made from seeds and nuts.

What about cooking oils? Yes, health stores carry peanut oil, safflower seed oil, sunflower seed oil, cottonseed oil, corn oil, linseed oil, and many more. You may use these oils for cooking and also upon raw salads. It is true that these oils have depleted enzymes—but enzymes in your raw salads, and your digestive enzymes, will combine with the unsaturated fatty acids from these oils to build good health. It is all like an enormous chain link. Each link fits in to the next and forms a whole. Take away one link, the entire chain falls into disuse.

Seeds and nuts have been created by Nature to become a vital link in your body chain. Use seeds and nuts in all forms to strengthen your body chain.

POINTS TO REMEMBER IN THIS CHAPTER.

1. Seeds are rich in all known nutrients as well as enzymes, having been placed there by Nature. Eat *raw* seeds. That is, seeds which come in the shell and have not been contaminated with by harsh chemical actions. Daily, eat one handful of seeds which you shell, yourself. Sprinkle seeds on salads, in soups, or on vegetable salads. Add seeds to yogurt and other beverages.

2. Toast your own seeds, for a different taste, but balance your diet with raw seeds and toasted seeds.

3. Nuts are prime sources of vitamins, minerals, unsaturated fatty acids and enzymes. Select unshelled nuts for enzyme power. Shell them yourself. Keep a bowl of nuts on a table, together with a nutcracker.

4. For variety, try seeds and nuts in such forms as oils, butters, spreads, flour, pulverized to be sprinkled on salads, soups, desserts, stirred in a beverage, hot or cold. Use seed and nut oils in cooking.

5. Above all—remember that the *raw, untreated* seed or nut is the highest source of enzyme power. Chew them well to release enzymes and enjoy better health.

13

HEALING POWERS OF
RAW JUICES

Raw fruits and vegetables, and their freshly squeezed juices, contain so many enzymes as well as other vital nutrients, we often wonder if life could exist if Nature did not give us these foods. The first command issued by God to Adam was for him to eat the fruits of the trees of the Garden of Eden. The very sustenance of life depends upon the nutrients you find in these raw foods.

Historically, we read that Moses exempted fruit farmers from military duties. Ancient pagans planted the olive tree to Minerva, the date tree to the Muses, and Bacchus received the grape and fig tree in special homage.

Ancients may have known little about technical, scientific data with regards to enzymes in these fresh fruits, but they did see what was known as "magic healing properties" and looked to fresh fruits and vegetables for building life and health.

When Noah left the Ark, following the Great Deluge, he lost no time in planting a vineyard. Joshua sent emissaries to the distant land of Canaan who reported back to him that the gardens were laden with fruits and vegetables. This indicated that the dwellers of Canaan were strong and healthy. Ovid, the Latin poet, speaks of a Grecian Golden Age when fruits were eaten regularly. It is said that much of the health attributed to the Greeks was possible by the copious eating of fruits and juices.

The legend of Prometheus who first stole flames from heaven, tells us that there was a healthy time in history when fruits and vegetables were not cooked, but eaten raw. In fact, there was a sect of India known as Gymnosophists, which existed nearly entirely upon raw fruits and green vegetables. The Gymnosophists maintained that whatever was eaten had to be sun-ripened and prepared by Nature. Understandably, we can see that raw fruits and vegetables are hailed as miracle healing foods. Why? Undoubtedly, because their enzymes are in full functioning force, not having been destroyed by processes of cooking.

What are fruits?

Botanically speaking, fruits are the edible portions of plants which result from the growth of pollinated flowers. The plant prepares the soft, delicious pulp of the peach, plum, apple, pear and orange, etc., for eating purposes. The seed, over which the fruit grows, is intended for plant reproduction. Of course, seeds are chock full of enzymes and nutrients and may be eaten, too.

Fruit enzymes favor digestion.

Enzymes create *levulose*—natural fruit sugar. This is prepared in fruit, by enzymes, in a state of complete absorption and assimilation into your bloodstream. When you eat other foods containing carbohydrates, your digestive enzymes must change that starch into glucose, a basic sugar, which is then absorbed and utilized by your digestive system. Even after absorption, some of this glucose is again enzyme-converted into glycogen to be stored in your muscles and liver to be used as body fuel when needed. Then, if you need glycogen, your enzymes must "burn" it to change it back into usable glucose sugar. But this is not true of fresh fruits which have enzyme pre-digested levulose. Your system *instantly* absorbs this levulose to give you powerful energy and bursting vigor.

Healing powers.

Juices are valuable to relieve hypertension, cardiovascular and kidney diseases and obesity. Good results have also been obtained in rheumatic, degenerative and toxic states. Juices have all-around protective action. The high buffering capacities of the juices reveal that they are very valuable in conditions of acidosis.

Raw fruits and their juices are a must! The living cell portion (enzymes) in fruits are destroyed by heat. You need to take raw fruits and drink juices in an uncooked and untreated form.

Enzyme Valuable Raw Fruits

Apples.   Rich in iron, minerals and enzymes which help in the metabolization of fatty foods. Enzymes spark function of maltic acid which heals internal inflammations.

Cocoanuts.   Grown close to the sea, rich in minerals which are used by enzymes to heal stomach and liver distress.

Lemons.   Excellent vitamin C source which is used by enzymes to build capillary and skin cells. Enzymes unite with bio-flavonoid substances in the lemon to neutralize excess acidity in the stomach.

Oranges.   Rich in minerals and blood-building iron. Vitamin C tones up blood vessels and promotes mental vigor.

Peaches.   Enzymes use the minerals of this fruit to build skin health.

Plums.   The highest concentration of vitamins A, B-complex and C are located in the meat close to the surface. Enzymes extract these nutrients and the minerals of silicon and sulphur to beautify hair and skin.

Pineapples.   Bromide is activated by enzymes to parallel and aid in hormonal secretions of the pancreas. A special enzyme, rich in the pineapple fruit, aids in overall food digestion and assimilation.

Strawberries. High iron content which is used by enzymes to improve your blood, strengthen your liver, build resistance against catarrh and normalize the glandular system. Has valuable vitamins and minerals.

Grapes.   Rich in iron and have a marvelous internal cleansing quality.

Figs.   The seeds have an undefined substance which enzymes appear to use to aid in overcoming constipation. Enzymes change the minerals into a substance that has been compared in nutritional value to human milk.

Berries.   All are rich in calcium and iron and needed by enzymes to build a rich blood supply.

Fruits help pregnancy.

"Fruit is like a medicine to the kidneys," states Owen S. Parrett, M.D., in *Life and Health*. "Last week I delivered a small woman patient of triplets. The total weight of the three was 21 pounds, 9 ounces. This patient showed albumen before pregnancy. Since multiple births predispose to toxemias of pregnancy and she showed increasing albumen, I asked myself whether she would carry throughout without losing her babies or jeopardizing her own life.

"Fruit was the answer, and she came through with no complications. Fruits and juices are kind to the liver and kidneys, which bear the brunt of childbearing. I cut out meat and eggs from her diet, giving her some milk, cottage cheese and *heaps of fruit and juices.* The three lads were big and husky and any one of them would do credit to his mother."

Dr. Parrett emphasizes, "Many people find themselves waking in the night because of what they ate for supper. Fruit will let you sleep."

Ulcers.

Famed Garnett Cheney, M.D., of the Department of Medicine at Stanford University treated 65 patients who suffered from ulcers. His treatment was simple—every day, they drank one quart of freshly squeezed cabbages, for a period of from 6 to 12 weeks. Within 5 days, all but three of his patients had relief from ulcer symptoms. At the end—nearly all were cured. Dr. Chenny suggested that the juice be squeezed right to the pulp and immediately used. It must not be heated. Apparently, the unidentified enzyme in the raw cabbage juice will lose its ulcer-healing powers by heating. You may add some fresh celery or carrot juices if you dislike the pure cabbage juice taste.

Arthritic conditions.

Gout is an arthritic condition in which the blood has an excess of uric acid. Other arthritic symptoms occur. Surprisingly enough, raw cherry juice was used for such an ailment with astonishing success. Ludwig W. Blau, M.D., in *Texas Reports on Biology and Medicine* (Vol. 8) tells how he treated a dozen cases by giving them freshly squeezed cherry juice as well as raw, uncooked cherries. "No attacks of the gouty arthritis have occurred on a non-restricted diet in all twelve cases, as a result of eating (the equivalent in juice form) about one-half pound of fresh cherries per day." Dr. Blau describes further that his cases responded best to such cherries that were sour, black, Royal Anne or fresh Black Bing. Enzymes in the cherries and their juices appear to have some influence in relieving arthritis.

About Vegetables

The raw vegetable is the second food that is rich in enzymes. What is a vegetable? Turning to botany, we see that it is a plant cultivated for its edible portions. Nearly all vegetables are surrounded by a strong outer skin that acts as a barrier to prevent escape of the essential food elements such as enzymes. How do these miracle workers in raw vegetables and freshly squeezed juices help build your health? Read on:

Alfalfa.	Rich in minerals, enzymes use these substances to help maintain a healthful acid-alkaline balance in your body. Enzymes cause amino acids to regenerate body tissues.
Artichoke.	Enzymes work to hydrolize natural insulin into levulose to give you a natural energy power. Also beneficial for those with poor digestive abilities.
Asparagus.	Enzymes use the mineral and vitamin content to give you a healthy blood supply and visual strength.
Beet.	Well known for blood-building powers; enzymes stimulate the lymphatic flow throughout your circulatory system by means of the minerals in the beet.
Broccoli.	Enzymes use these vitamins and minerals to keep a healthy body water balance.
Cabbage.	An unidentified enzyme in cabbage was instrumental in treating, halting growths and coping with ulcers as well as easing duodenal disorders.
Celery.	The vitamins in celery aid in emotional disorders; enzymes need celery nutrients to help your body cast off carbon dioxide waste substances.
Endive.	Enzymes produce a natural bowel movement, soothe a distressed stomach and liver.
Kale.	Valuable Vitamins A and C build skin, hair and eye health.
Lettuce.	The minerals are utilized by enzymes to ease conditions of excess stomach acidity as well as difficult bowel movements.
Mushroom.	B-12 is assimilated by means of enzymatic action to enrich the bloodstream.
Mustard Greens.	A potent source of skin building Vitamin C.
Okra.	Low carbohydrate count makes this a good vegetable for reducers. Okra has some iron and calcium which is used by enzymes in conditions of colitis and intestinal disorders.
Onion.	Unhappily, this is not a popular vegetable because it is anything but fragrant; yet, enzymes need certain nutrients in the onion to stimulate natural functions of the gastric tract.
Peas.	All peas have ample supplies of protein and minerals.
Peppers.	Enzymes especially need the silicon in peppers to give

you skin, hair and fingernail health; in some persons, acne and blemishes clear up after enzymes attack them, fortified by substances in raw peppers and extracted juices of green or red peppers.

Radishes. Soothe your nervous system and stimulate enzymatic flow in your digestive system with the high sulphur supply of radishes.

Tomato. Best source of B-2 and C as well as amino acids. Enzymes take these amino acids and help you neutralize excessively acid stomach conditions.

Raw vegetable juices.

Feed yourself an enzyme cocktail by taking a glass of any raw vegetable juice, singly or in whatever combination you prefer. The phosphates of succulent vegetables, where nourishment is mostly in their juices, are of course soluble. They contain the enzymes which you need. You see, since vegetables are surrounded by fibers that are protective and tough, it means that within these fibers are found highly potent sources of enzymes. When you put raw vegetables into a juice extractor, you split these fibers and the rich enzyme content is released into the juices. If you chew vegetables, remember our earlier admonition—*chew, chew, chew.* You must get right down to the locked in nutrients of both fruits and vegetables and this is possible only by chewing or by drinking the freshly extracted juices.

Importance of juices.

Freshly squeezed juices serve to cleanse your internal system, rejuvenate, repair and rebuild your body by means of their powerful enzyme content. Within the cellulose fibers are many vital enzymes that may be difficult to extract by ordinary chewing.

Furthermore, it has been noted that within 18 minutes after you drink a glass of a raw juice, the liquid is digested, assimilated, absorbed and speedily used to nourish and regenerate your cells, tissues and organs. And all of this is performed with as little effort as possible on the part of your digestive system. If you have stomach trouble, or if you have indigestion, look to the values of raw juices, both fruit and vegetable.

A famous nutritionist says of fruits—They are the finest source of energy available to us, requiring almost no digestive effort, leaving no chemical residue to clog the body machinery when it is burned for fuel.

It gives a certain soft bulk to the bowel, in contrast to meat which is mostly putrefactive.

Preparing juices.

Obtain an electric juice extractor. The household goods section of any reasonably large department store will have different types of electric juicers for sale, at different prices. All are comparatively simple to operate. Follow the instruction booklet that comes with an extractor. If there is none, ask the store to get one for you, or ask for the name and address of the manufacturer and write him directly.

When making juices, use freshly purchased raw fruits and vegetables. Wash carefully with a special brush (for vegetables) under running cold water. Wash fruits by rubbing your hands on them beneath running cold water. Cut the fruits and vegetables into pieces than can easily be inserted into the feeder of your electric juicer. Close the lid if your juicer has one, flip the switch and presto—you have a cup or a quart of freshly prepared juices. Because juice making is simple, arrange to squeeze some every single day. Always clean your juicer, as per instructions, after use.

You may also inquire at your local health food store for juice extractors. Many such stores have different varieties so you have a selection to choose from.

If you find it necessary to store your juice, put it in a jar, screw on a tight cap or lid, then place in your refrigerator immediately. This keeps the juice cool, preserves its enzymes and other natural nutritional content.

Sample Juice Recipes

Beauty Beverage. Mix together equal portions of peach, pear, apple and watermelon juices.

Raw Vegetable Mineral Broth. ½ cup freshly squeezed celery juice; ½ cup beet juice; 3 tablespoons sauerkraut juice; 3 tablespoons parsley juice; 1 tablespoon radish juice; 2 tablespoons cabbage juice. Serve after stirring vigorously.

Appetite Tonic. 6 tablespoons green pepper juice; ½ cup tomato juice; 2 teaspoons onion juice; ½ cup lettuce juice; 1 teaspoon lemon juice; Combine together in a cocktail shaker (or blender), then add 1 tablespoon horseradish juice. Stir vigorously. Sip slowly one hour before you have to eat.

Morning Eye-Opener. ½ cup orange juice; ½ cup grapefruit juice; 5 tablespoons strawberry juice; 4 tablespoons peach juice. Blend together, stir in 2 tablespoons unflavored gelatin. Mix until all ingredients are thoroughly assimilated. Drink before breakfast.

Combinations are limitless. As you become more familiar with your juicer, you will learn there are many different types of flavorsome mixtures, depending upon individual tastes. Mix together any juicy fruits to add whatever color you prefer; do the same with juice-giving vegetables. It's fun . . . and deliciously healthy. Always be aware that in freshly squeezed juices, you have the greatest source of enzymes.

Not only will you taste the difference in fresh raw fruits and vegetables and their extracted juices—but you'll feel the difference, as well!

When to drink juices.

Digestive enzymes seem more active if you drink a glass of freshly squeezed juice before each of your meals. Before breakfast, a glass of freshly squeezed fruit juice; before lunch, alternate between vegetable and fruit juices; before dinner, vegetable juices.

If you do not have an extractor, then remember to chew, chew and chew your fruits and vegetables.

You may care to purchase cans of raw juices from a local health store. These have been prepared under the most exacting conditions and the precious enzyme content have been preserved to a large extent. Many health stores have juice bars—you can order a drink and watch it being squeezed while you wait.

Home Remedies with Fruits and Vegetables

Considering the power of enzymes in raw fruits and vegetables, it is not surprising to hear of quite a number of home remedies for various ailments by means of using these natural foods. Here are a few that I have come across. Nearly all call for uncooked raw fruits and vegetables; some call for cooking so it is apparent that other nutrients are able to bring about relief and even cure of the ailment.

Throat Disorders. Try taking two teaspoons of raw lemon juice before you eat a meal.

Poor Digestion. Sprinkle lemon juice on a raw vegetable plate and keep it in a refrigerator for about 30 minutes prior to eating. The juice will partly digest the tougher parts of the plants by means of its enzymes, rendering the vegetables more digestible for you.

Foot Troubles. The Chinese rub sliced lemon over any painful area.

Country folk who have to walk long distances will start by bathing or washing their feet in water diluted with much lemon juice. A corn is often treated by fastening a lemon slice over the area and allowing it to remain the whole night.

Bruises. A bruised or injured leg, wrist, etc., may benefit by having a banana peel wrapped around; the inner part is rich in enzymes that soothes inflammation, burns, scalds, swellings, simple wounds, sores, neuralgia. It is usual to bind the inner surface of the peel to the affected part and renew every few hours.

Inflamed Parts. A thumb or toe (or any body part) that is inflamed and swollen could be alleviated by a raw potato pulp poultice. "Oldtimers" apply it is a dressing, renewing it every night and morning.

Skin Trouble. Boil potato skin in water, use the water as a beverage—three tablespoons freely, at any time. Incientally, if you have a burn, apply a mash of scraped raw potato for soothing poultice.

Stomach Help. Place apple peelings in a flameproof glass pot. Cover with water. Bring to a boil. Simmer for three minutes. Strain off the water. Add buckwheat honey. Drink before and after meals to help strengthen and soothe stomach functionings.

Sore Eyes. In some country districts, over-ripe apples are used as a poultice for sore eyes. Apply the pulp over closed eyes, hold in position with a bandage for 60 minutes. Then remove.

Rheumatic Pains. Boil apples to a jelly and apply as a liniment; that is, rub into the affected parts freely at any time.

Skin Beautifying. Hollywood should try this for a complexion of peaches and cream. Massage your face with the inside of a peach or apricot, nightly, for 5 minutes. Do not rub away the moisture. This will cleanse your skin, free the pores; the enzymes in the peach or apricot will tighten your muscles, help prevent sagging tissues and melt lines. Rub this fruit over any sore or roughened body part to cause regeneration.

Skin Tonic. Cucumber has an enzyme that makes a wonderful skin tonic. Take three ounces of cucumber juice, mix with one teaspoon of glycerine (sold at any pharmacy). Massage this into your face at night after washing. This skin tonic removes certain spots, blemishes, sunburn effects, irritation; in some cases lightens freckles.

Hemorrhoids. Pulped carrot has been seen to relieve conditions of hemorrhoids. Apply either warm or cold and renew every three hours. Prepare it as a poultice.

Kidney and Bladder Distress. Take a handful of the fresh tops of carrots, put in a pot. Cover with boiled water. Allow to stand, for three hours. Drink the juice before meals—a glassful, three times daily.

Sore Throats, Nervousness. Place the green tops and tough stalk parts of celery which are generally discarded, into a pan. Cover with cold water. Bring to a boil. Simmer gently for 10 minutes. Strain. Drink one glass of this liquid before meals, three times daily. May be taken hot or cold.

Healing Wounds. Parsley is cited as good for bruised, inflamed and lacerated wounds. Apply to the wound; also apply to parts of the body that have received bites and stings of insects.

Eye Fatigue. Take a handful of the freshly plucked leaves of tomatoes, place in a pot, cover with hot (not boiling) water. Allow to stand for 15 minutes. Strain. Drink two tablespoons of this beverage before meals, thrice daily. It is claimed to be a wonderful tonic for tired eyes and strained optic nerves.

Body Sores. The green leaves of the sweet-scented violet are most beneficial for body sores. Take freshly picked leaves, wash in cold water, put in a jar. Pour on boiling water. Cover the jar, let it stand throughout the night. Strain off the liquid in the morning. Drink one-half a cup every two hours. You may use this same liquid for skin sores as a local compress. Dip muslin into the solution, then place this over the sore, keeping it on for three hours. Renew regularly until the sore is healed.

Beauty Treatment. Make the violet tea (described immediately above) and dip a piece of cotton wool into it. Rub your face with this cloth for ten minutes every night. Then dry and if you desire, apply a natural skin cream.

Headaches, Toothaches, Neuralgia. Enzymes in the tops of beets help relieve these pains. The water in which both roots and tops have been boiled make an ideal application for such ailments. Rub this water over your forehead, on the back of your neck for toothache and neuralgia.

Dandruff. Massage the beet water into your scalp with fingertips, every night.

Pimples, Acne. Mix three parts of beet water to one part of white vinegar. Sponge the affected body parts with this mixture.

Stiff, Aching Joints. Enzymes in bran are most helpful for joint distress. Of course, boiled bran is enzyme-deficient but other nutrients rally to the cause. Pour one pint of boiling water over two tablespoonfuls of clean wheat bran. Boil for 15 minutes. Strain. Drink one cupful whenever desired. Minerals in the bran will ease tired joints. Or—heat up the bran, drain off and discard the water; use the heated bran as a poultice for painful muscles, spinal weakness and stiff joints. You may apply them in a muslin bag for convenience.

Tired Feet. Rub the soles of your feet with a garlic clove every night before going to sleep.

Arthritis. Sawdust, that is, pine sawdust, is an old remedy our pioneering ancestors knew of and used. Since sawdust is mineral rich, we can see how helpful it may be for painful bones which are probably starved for minerals. Obtain sawdust from a newly cut wood source. Scald with hot water to which has been added a little bran to better retain the moisture. Put in a muslin bag and apply to the affected part. Let it remain for 60 minutes. Renew regularly until pain has eased. If you cannot obtain fresh pine sawdust, then settle for any sawdust. Ask your pharmacy to get it for you. If he cannot, then look in the classified pages of your telephone book under Pharmacies —Homeopathic. If there are none, look for Pharmacies—Herbal.

Bronchitis. When I was a small boy living in the country, my mother would relieve any stubborn bronchitis cough in this way: she would soak stiff brown paper in vinegar and then sprinkled one side with ordinary pepper. This was applied to my chest; the peppered side in contact with my chest. I slept this way all night. The results were wonderful.

IMPORTANT POINTS OF CHAPTER 13.

1. For potent enzyme action, eat as much raw fruits and vegetables as you can.

2. Drink freshly squeezed raw fruit and vegetable juices.

3. When making juices, use freshly purchased foods. Wash carefully, cut into small pieces, insert into your juice extractor. Drink in between meals or even as a nightcap.

4. Try some of the sample recipes included in this chapter; make up your own, depending upon your individual tastes.

5. External application of enzymes is possible with some of the home remedies at the end of the chapter, for a variety of ailments.

14

HOW TO COOK AND PRESERVE ENZYMES IN FOODS

Cooking vs. enzymes.

We are told by enzymologists that there are two types of these miracle life builders. *Endogenous enzymes* are those produced within your own body. *Exogenous enzymes* are in raw foods. The more exogenous enzymes you consume by eating raw foods, the less endogenous enzymes will be needed; this means less strain and wear and tear on your internal tissues. In other words, by increasing your raw food intake, you will be feeding yourself valuable enzymes, thereby making up for any deficiency within your system and providing much needed relief for those overworked enzymes that you already have.

But, we are also told by enzymologists that raw food (exogenous) enzymes are highly perishable. They cannot withstand hot temperatures such as those used in cooking or any processing method. Whether you bake, roast, stew, broil, boil, fry, you will kill these enzymes. Yet, it is extremely unpalatable to eat raw meat, fish, potatoes, not to mention indigestibility. So—you do need to cook those types of food. What to do?

As explained previously, all fruits should be eaten raw! I do not know of any fruit that has to be eaten in a cooked state. This means you eat raw apples, pears, peaches, plums, grapes, berries, oranges, grapefruits, etc. You may drink their freshly squeezed juices, if you prefer and this, too, gives you a wonderful supply of food enzymes.

As for vegetables, many of them may also be eaten raw. Some are even more delicious when raw; try a fresh raw turnip and you will never want to eat one that has been cooked. You may enjoy raw shredded beets, chopped cauliflower, brussels sprouts, chard, kohlrabi, etc. The list is endless. Eating raw vegetables will give you the necessary raw food (exogenous) enzymes that you need.

What about cooking those vegetables which you cannot eat raw? A temperature of about 212° F. will destroy enzymes. You may go as high as 122° F. and some enzymatic life may remain. Yet, you may be wondering how to cook vegetables and still retain as many enzymes as possible. And, for the sake of variety, you may want to serve a dish of cooked apples, stewed prunes, or even a baked peach pie. How can enzymes be retained in cooked fruits?

The answer is that a few cooking methods will preserve appreciable amounts of enzymes. Other cooking methods will *not* save enzymes; so, if you do like to eat a baked peach pie, make sure that at the same meal, you eat a plate of raw fruit slices to make up for the loss.

Waterless cooking.

This is the best way of preserving as many enzymes as possible in cooking vegetables. Fresh vegetables have from 70 to 95 percent water content which is enough for cooking them if you control the heat so that no steam escapes. The waterless cooking utensil should have a tight-fitting lid. The heat must be evenly distributed to the sides of the utensil and the lid. This cooks the vegetable by heat from all directions.

You do not add any water when cooking vegetables by this method; however, add two tablespoons into the preheated utensil in order to replace oxygen by steam. This prevents enzyme dissipation and loss. You must remember that the heat must be kept low for the first five minutes so that no steam will escape. If you cook raw vegetables in this manner, you will be preserving a good amount of the natural enzyme content.

Oriental "steaming" method.

For centuries, the Chinese have cooked their vegetables in a surprisingly simple, yet enzyme-rich way that is also most appetizing. They heat a pan; when very hot, they add about one teaspoon of vegetable oil which becomes heated in a matter of a moment or two. Then they drop their vegetables—cut in tiny pieces, into the pan. Stir constantly at high heat for just one minute. Remove and serve immediately. The vegetables should be crisp and young; if the process is done properly, the vegetables will be crunchy and brightly colored. The trick here is that steam has tenderized the vegetables, not water. Enzymes seem to be retained to a large degree when vegetables are cooked this way. The utensil, incidentally, should be of heavy gauge.

Braising or panning method.

This method, slightly different from the above, requires very little water; what you see is the steam formed from the vegetables own juices. The liquid becomes a part of the flavored sauce which you serve with the vegetable. Shredded cabbage, kale, spinach, okra, and snap beans are some of the vegetables cooked successfully by this method.

Cut the vegetable in small pieces and drop into a pre-heated heavy pan on top of your range. Add just a teaspoon of any vegetable oil to prevent sticking. Put on a tight cover to hold in the steam. (This latter is important as evaporation means loss of precious nutrients.) Cook over low heat until just tender. Serve immediately. This method conserves a lot of enzymes and many of the vitamins, minerals, etc.

Home-type steamer.

Steaming vegetables in a home-type steamer, consisting of a perforated pan placed over rapidly boiling water, may take slightly longer than by boiling. There may be some color loss and also depletion of enzymes and nutrients. Vegetables with more cellulose fibers may be cooked this way.

Avoid salt or soda additives. Mrs. Weldon was a local celebrity in the small rural upstate New York village where she resided. County fairs, church benefits, charity bazaars all awarded her blue ribbons when it came to cooking. She could prepare a dish that would be the inspiration for an artist insofar as color schemes are concerned.

"My cooked vegetables look like the colors of the rainbow," declared Mrs. Weldon when she was given a special gold ribbon award by the judges of a tri-county award. "Did you ever see such green and yellow colors?"

A runner-up had to admit that Mrs. Weldon's vegetable pie entry was superior in appearance and in taste, too. The runner-up, Mrs. McCann, was only slightly envious as competing country cooks can be. "I do wish you'd let me watch you cook, Mrs. Weldon. Maybe I'll find out your secret."

Mrs. Weldon laughed, linking arms with Mrs. McCann. "Sorry. My secret is one that was passed on by my mother to me, and I'll pass it on to my daughter." Then she saddened. "Even though my mother may have known a special way to cook vegetables, it didn't help her health, poor soul. Always sick and wan."

Mrs. McCann was well aware of the health of the Weldon family. Her husband, Dr. McCann, treated the senior woman, then this one and her daughter, as well as other members of the family. They were always sick and ailing, among the first to contract any contagious disease, the last to survive.

"I wish the Weldons would look as good as Mrs. Weldon's cooking," was Dr. McCann's frequent lament.

Only after much pleading, did the prize winner let Mrs. McCann watch as she cooked vegetables. But only after she extracted a promise that the "secret" must not pass her lips. "Anyway," Mrs. Weldon said as she started cooking asparagus and potatoes, "you folks look healthy, fit as a fiddle. Not like we Weldons—always ailing from one thing or another. Maybe you'll let me in on *your* secret of keeping healthy."

Mrs. McCann watched the prize winner as she put vegetables in a large pot on top of the range, poured cold water over them, put them to a high heat temperature and just as the water began to boil, added one half teaspoon of salt and a tablespoon of baking soda.

"See?" the prize winner winked at Mrs. McCann. "The real secret is in the salt and baking soda. That's how the flavor and color are brought out."

Mrs. McCann was perturbed. "So that's how you do it!"

"You promised not to tell anyone!"

Mrs. McCann laughed. "Never fear. I wouldn't dare tell anyone to cook vegetables this way. They'd only get sick. No wonder you folks are always ailing. Baking soda, salt or any of those artificial flavoring agents destroy whatever vitamin and enzyme content you have in the vegetables. You're eating lifeless food—and becoming lifeless yourself."

Mrs. Weldon, the champion prize winner, was stunned. She had heard some "foolish talk" about avoiding soda and salt, but used them because "they make the vegetables look as pretty as a picture."

There was no disputing that fact. "But the picture is dead, Mrs. Weldon. You're eating dead food—and that explains why your mother and her family were always ailing, and why you and your family aren't in the best of health. You need more vitamins, minerals, enzymes—and you're destroying them by this bad cooking method."

It was not easy to convince Mrs. Weldon that she was wrong. After all, wasn't she a champion prize winner in just about all major cooking contests in the entire area? But after much persuasion, she decided to change her methods. For a period of eight weeks, she would use no salt or baking soda or any additive when cooking vegetables. If her health, and that of her family, improved, she would concede her error.

Seven weeks later, the little area was swept by a bout of the Asian Flu. A year before, the same epidemic occurred and the Weldons were the hardest hit and the most to suffer. Dr. McCann observed, "They're so weak, their resistance is at a low level."

But this year, they were the last to be infected by the flu, the first to recover. Added to this joy was the zest and energy the whole Weldon family experienced. Life was really good.

"I'll bet it's because I stopped using salt and soda," admitted Mrs. Weldon. "Maybe the vegetables don't look as good, but my family feels better and that's what really matters."

Why was this so? The answer is that salt draws moisture out of the foods; therefore, when a vegetable is subjected to salt at the start of cooking, its juices, rich in vitamins, minerals, enzymes, are drawn out and dissolved. Baking soda does the same and while it may intensify vegetable color, it also depletes nutrient store—and kills enzymes.

Green vegetables, in particular, contain plant acids which react chemically while cooking and combining with the coloring matter. A prolonged and over-soaked cooking causes loss of the acid as well as coloring. (Not to mention nutrient depletion.) This explains why so many people maintain that they eat copiously of cooked vegetables but are still ailing (as with Mrs. Weldon above). They are eating lifeless vegetables and being denied health-building nutrients and life-giving enzymes.

Danger in hot foods.

Spill a hot drink on your hand and you are burned. Well, what happens when you drink a scalding liquid! Yet, there are so many who must sip boiling hot beverages and gasp with the first swallow since the fiery pain is hardly bearable. They just cannot drink any beverage unless it is really boiling!

Just as heat in cooking is destructive to food enzymes, so will excessive heat in drinking liquids be destructive to those enzymes within your system.

A recent study was conducted on this subject of enzyme destruction in the body because of hot liquids consumed. It was found that the intake of drink at high temperatures is positively associated with mucosal abnormalities in the stomach. A team of British doctors (*Lancet*, 9:56) interviewed 155 patients who had had successful gastric biopsies.

At the conclusion of the interview, each was given a fresh cup of boiling tea. When the patient drank the tea, its temperature (of the tea) was taken. The results were these—only 2 out of 13 patients who drank tea below 122.5° F. showed any gastric enzymatic abnormalities. But those who drank tea above 137.5° F. did show disorders at the rate of 14 out of 18.

It is conclusive that overly hot foods that are consumed will exert a harsh, destructive effect upon your digestive enzymes.

Another survey reveals that a majority of patients with gastric cancer admitted that they liked foods that were hotter than those of the rest of their families. No doubt, the heat destroyed valuable digestive enzymes and predisposed a condition toward cancer.

Be patient. That's right—be patient and take your time. Don't bolt down a hot beverage and then rush to work or to the corner store or PTA meeting. When you're in a hurry, there is the familiar temptation to gulp down foods. Just as you deplete enzyme intake by improper chewing, you will destroy your own enzymes by bolting down very hot food!

It just takes a few moments until the hot food becomes more comfortable. Excessive heat from hot foods has a paralyzing effect upon your tongue's taste buds. When you cannot taste properly, neither can you chew properly and starch foods are denied mouth enzyme action because you swallow quickly.

If a tea cup is too hot, or if a bowl of soup or chowder is too steaming, wait a moment or two. Stir the food with a spoon or fork. Air being stirred into the dish will carry off some of the heat.

It has also been found that excessively hot foods cause internal complications which lead to ulcers, cancer, mouth, stomach and throat disorders. Destruction of enzymes because of such heat is undoubtedly a large causative factor.

When Buying Vegetables

Here are some pertinent points with regards to buying vegetables. The enzyme content is highest in good, fresh vegetables so start out right by being a wise consumer.

1. The vegetable should be fresh, firm, tender, with no bruises or imperfections.

2. Head vegetables should be obtained if there are only a few waste leaves. Leafy vegetables should not be wilted.

3. Use vegetables as soon as you bring them home. When allowed to stand, there is oxidation of valuable nutrients and enzymes. If they must be stored, wash them thoroughly, put in a closed container, in the middle shelf of your refrigerator to remain there until ready for preparing.

4. Wash vegetables well. Hold them under cold running water and use a scrub brush when necessary. This is important since insects, spray residues, dust, etc., cling to leaves and stalks of vegetables.

5. Cook whatever vegetables have to be cooked with the skins on. If you pare, remove as little of the outside skin as possible. The less you peel, the less you cook, the greater the nutrient store.

6. A good way to preserve enzymes is to wash leafy greens quickly but thoroughly (never leave them soaking in water), shake off water, dry quickly by whirling in a wire salad basket or cloth bag. Store in a hydrator. It should be fitted with a rack, beneath which is kept a fraction of an inch of water to supply continuous moisture. The vegetable and water do not touch. Fit the lid tightly. (Any department store or housewares outlet will sell you a hydrator. Ask for waterless cookers at the same outlet.)

7. As for cooking, remember to use any waterless method as the main technique. Use vacuum-sealed, stainless-steel cookware for best results and greatest enzymatic power.

What about fruits?

Much emphasis has been placed on vegetables. Why not fruits? Because fruits (with an occasional exception) may all be eaten raw and that is the way you should eat them for enzymatic health. You may cook fruits, make applesauce, pies, cakes, puddings, etc., but you will not have any enzymes this way. Go ahead and eat cooked fruits but be sure to eat plenty of fresh, raw fruits every single day.

Organic edibles.

An important tip is to buy fruits and vegetables that have been organically grown, and are garden-fresh, without contamination from artificial chemical fertilizers or poisonous sprays. You can find such organic farmers by asking at your local health food store. Or, write to your State Agricultural Department, Extension Division, and request the names of farmers who do not use any chemical sprays or fertilizers. Certain magazines carry advertisements of people who can supply you with foods that are pure and organically grown: *Organic Gardening*, Emmaus, Penna. *Prevention Magazine*, Emmaus, Penna. *Natural Food and Farming*, Atlanta, Texas. You may buy many products through the mail, with confidence, from advertised companies in these magazines.

Your Guide to Better Health

1. Taste is not a reliable guide. Remember, all food that you eat should build health, not tear it down. Eat foods that are nutritious and healthy for you—this means all natural foods and pure foods.

2. The so-called refined or "enriched" foods have less nutrients and negligible enzymes than natural ones. Purchased foods should be free from chemical additives, sprays, preservatives, etc. Organically grown foods are the best.

3. Raw food is your watchword. Cook as little as is necessary and by the methods described earlier in this chapter.

4. If possible, try to grow some of your own foods in a backyard garden patch. Learn how to be an organic gardener-farmer. The magazine, *Organic Gardening*, Emmaus Penna., is the Bible of natural method gardeners.

5. Before you plan a meal, ask yourself: Do I have to cook this food? If so, can I use a waterless method? Each meal must have a dish of raw vegetables.

6. When someone chides you for "bothering" with fresh, uncontaminated food which is difficult to obtain and costs a bit more, remember this: "With food at such a high price, how can you remain healthy? It may cost you more to buy better food, but it costs even more to go into a hospital, not to mention all-time high doctors' prices. So—how can you afford *not* to be healthy?"

About enzyme powders and tablets. Suppose you feel you would like to take enzyme supplements, just as you may be taking vitamin, mineral, protein and any other food supplements. Are there enzyme supplements? Indeed, yes.

These are exogenous enzymes that are extracted from foods and cells, then dried to a powder. Concentrated into a powder or tablet, by means of a special process at blood heat, a temperature that will not destroy enzymes, you can take such supplements to replace any shortage in your system.

These powders and tablets are available under many different brand names at nearly all health food stores. You may also inquire at your local pharmacy for any such enzyme supplements.

CHAPTER 14 IN A NUTSHELL.

1. There are two types of enzymes—endogenous enzymes which are produced in your body. Exogenous enzymes which are found in raw foods.

2. Raw food enzymes are perishable and will be destroyed by hot temperature cooking.

3. Vegetables which must be cooked should be done so by means of waterless cooking, oriental "steaming" method, braising or panning method.

4. Avoid salt or baking soda when cooking any food.

5. Foods should be *comfortably* hot, not scorching or scalding when you eat them.

6. Purchased vegetables should be fresh, free from bruises or blemishes. When washing, be thorough. Remember—rinse and wash but don't soak in water!

7. Select organically grown foods as much as possible.

8. Fruits should be eaten raw. For variety, eat some cooked fruits, but don't do this at the loss of raw fruits.

15

GROWING YOUNG WITH ENZYMES

Suppose there could be one food that would supply you with just about all the necessary enzymes you need to build and maintain youthful health and vigor. Wouldn't that be a wonderful miracle of Nature? The truth is that no single food contains all elements your mind and body require. Nature has seen fit to provide you with a balance of different edibles to be eaten raw and cooked as a means of giving you a treasure of nutrients that give you life. Of course, even if one single all-purpose food did exist, it would soon become tiresome, existing just upon the same item all the time. Nature gives you a variety so your appetite will always look forward to a new adventure in good eating and good foods.

A miracle enzyme food.

But—there is *one* food that is the richest source of enzymes. Other valuable substances in this food have been seen to influence function of vitamins, minerals, proteins to perform a better job in keeping you healthy and young. This particular food has been known to the Europeans for decades and is surprisingly new in our country. You are, undoubtedly, aware of the long-living Bulgarians who include yogurt (a fermented milk rich in enzymes) as part of their staple diet. Many have maintained that substances in yogurt help prevent premature greying and loss of hair as well as tooth decay. Yogurt's enzymes help in the assimilation of calcium to build strong bones and teeth and also aids in the body's own manufacture of the valuable B-complex vitamins; further, yogurt is a prime source of pre-digested protein.

However—there is still another food which has escaped attention among modern researchers. This food is a fermented milk edible which is delicious, enzyme-rich and about the *only* food that we may safely say is a fountain of youth.

The food which contains the richest known source of enzymes and other substances which will spark function of digestive enzymes is known as *KEFIR*. It is highly superior to yogurt as we shall now discover. Kefir (sometimes called *kephir* or *kifir*) is a grain food, originating in the Caucasus, which is placed in milk, allowing it to sour and then eaten. Because kefir grains are prepared and used under circumstances which do not destroy its dynamic-enzyme store, it is the most powerful source of enzymes that we know of today!

The Bulgarians have been using kefir for more centuries than we can deduce. The superior health of the robust men, women and children living in the wilds of the Caucasus may also be traced to kefir which is part of their daily diet!

Kefir is superior to yogurt because it has a very low curd tension which means that the curd breaks up very easily into extremely small particles and enzymes are thereby released and instantly sent to work in building life and health. The yogurt curd holds together or breaks into lumps so full enzymatic release is slower and not as great as with kefir.

This small particle size of the curd facilitates its absorption into your system because it presents a large surface for enzymes to work upon. It has been found that kefir stimulates the flow of salivary enzymes, will increase the flow of digestive juices (enzymes) in your gastro-intestinal tract, it will also stimulate peristalsis (a wave of muscular contraction) in your intestinal tract by means of enzymatic function. Constipation can be a problem of the past when kefir restores enzymatic power in your intestinal tract.

Several decades ago, a noted Danish dairy bacteriologist, Dr. Orala-Jensen urged for greater use of kefir by explaining that it had a higher nutritive value than yogurt because of its abundance of yeast cells which are needed by enzymes as living food.

Today, kefir is widely used in Scandinavia as well as the entire region of the Caucasus where it is known that people not only live to be well beyond the century age, but have youthful mental and physical functions. Enzymes in kefir are undoubtedly providing that mysterious "something" that may be the answer to finding a fountain of youth. And yet, here in the United States, kefir is relatively unknown.

How to make kefir.

To prepare kefir, you do not need any special equipment or conditions. You take kefir grains which are colonies of milk-fermenting yeasts and enzymes, place one tablespoon in a glass of milk, allow it to sour. (Usually, let it remain at room temperature, overnight.) When sufficiently thick and velvety smooth, eat with a spoon.

Jet enzyme tonic.

Here is how you can make a delicious, nutritious and health-building tonic that will put "jet action" into your entire enzymatic system. Into an ordinary drinking glass filled with certified raw milk, place two tablespoons of Brewer's yeast flakes, one tablespoon of kefir grains; stir vigorously until blended. Cover the glass. Let it remain at room temperature overnight. Eat this the next morning, about two hours after breakfast. You will discover that this Jet Enzyme Tonic will have a dynamic action in putting power into your entire digestive system.

Where to get ingredients. Why certified raw milk? Because pasteurization destroys or at least largely reduces valuable vitamins and minerals and precious enzymes. Homogenized milk has a greater destructive action upon the highly perishable enzymes. The only safe raw milk is called "certified raw milk" which means it has been produced under strict state and federal supervision; local medical milk commissioners also regulate manufacture of certified raw milk.

This milk comes from cows that are fed on food from green pasture, from a strictly inspected herd; milk workers also undergo rigid health examinations. Because certified raw milk often requires a prescription, you would do well to check with the Agricultural Extension Division in your state capital. Further information on where and how to order certified raw milk is available from the Walker-Gordon Certified Milk Farm, Plainboro, New Jersey. If you cannot obtain certified raw milk, substitute with soya milk made from powder which is sold at all health food stores. Mix with water and you have an enzyme-rich milk source.

Because kefir is so relatively unknown in the United States, you should ask your state Agricultural Extension Division where you may obtain these grains for making a special enzyme power-packed tonic. One mail order supplier is the R.A.J. Biological Laboratory, 35 Park Avenue, Blue Point, Long Island, New York. They have a mail order division for selling kefir grains direct to consumers.

Brewer's yeast flakes are sold in powdered form at just about any health food store or special diet shop.

Become the first in your neighborhood to discover the miracle health building powers of kefir—take the special Jet Enzyme Tonic every single day. Where all else may be inadequate, this Tonic will become a power source of natural and instant health by means of enzymes.

Getting the Most for Your Food Dollar

Since your life and health depend upon foods that you eat, you must spend a little more time in finding the best edibles. It is an investment second to none. Just as an engineer will invest time in searching throughout the world for materials with which to build a structure that will withstand the ravages of time, so should you—the engineer-architect of your mind and body, search for the foods that will keep you healthy and young. Here are suggestions:

Fruits: All fruits should be fresh, grown by organic fertilizer methods. Eat seasonal fruits, those which are tree-ripened. To eat out of season fruits, select the unsulphured, sun-dried, poison-spray-free fruits. Wash these dried fruits under running water; cover with water in a glass bowl. Let soak overnight or longer. Drink the juice and eat the fruits—without cooking.

Vegetables: Leafy greens are richest in vitamins, minerals and enzymes. Leaves should be dark and bright-green. Wash speedily and thoroughly. Whatever can be eaten raw, *must* be eaten raw. What has to be cooked, should be subjected to the least possible cooking time. Do *not* peel any vegetables unless they are too tough, bitter or rough and the peelings have to be done. Scrub vegetables well, but do not scrub off the skins which contain rich concentrated supplies of vitamins, minerals and enzymes. Serve vegetables as soon as possible after purchasing and preparing. Select vegetables that are organically grown.

Eggs: Rich in protein, you should eat eggs since they contain all essential amino acids. These same amino acids combine with enzymes to build strong body tissues and maintain health. Did you know that eggshell is a prime mineral source? If you can grind eggshell into a fine powder, you can use it for baking when combined with whole wheat flour. Enzymes are more abundant (as are other nutrients) in fresh eggs. Avoid frozen or dried eggs which not only taste leathery but have lost much of the valuable nutrient supply.

Eggs should be fresh, fertile; when you store in your refrigerator, place them with the small end down so the yolk is more evenly centered. To get the most benefit of the enzymes in egg yolk, separate it from the white; drop the raw yolk in a freshly squeezed vegetable juice combination and whirl in a blender for a delicious drink. You can also drop a raw egg yolk into any whole-wheat breakfast cereal, stirring vigorously. The result is a delicious flavor, a custardy consistency and a rich enzyme source.

Meats: Select government inspected meats. Glandular cuts are richest in nutrients; of course you will be cooking meat but you should use a low heat, broil or roast since this represents less loss of essential nutrients. Needless to say, enzymes will be destroyed by cooking—but other nutrients in cooked meats which are needed by your digestive enzymes as their foods, will remain preserved by this low heat cooking process. Organically raised meats are the best.

Fish: Prefer fresh fish which you keep refrigerated (not frozen) until time to use. Here is how you can tell that your whole fish purchase is fresh. Look for these signs: *Eyes*—bright, clear, bulging. *Gills*—reddish pink, free from slime or odor. *Scales*—tight to the skin, bright and shiny. *Flesh*—firm elastic flesh, springing back when pressed, not separating from the bones. *Odor*—fresh and clear, not offensive. *How long to keep fish.* Fresh fish is tastier when eaten right after purchase. After 4 or 5 days storage, there is a decrease in flavor even though it remains completely wholesome and nutritious. Buy fish in season.

Bread: Select bread foods that have been made with freshly ground, stone-ground whole-grain flour. Avoid bread made with bleached or "enriched" flours. Nutrients and enzymes have either been largely or completely destroyed.

Seasonings: Harsh condiments and chemically prepared seasonings are destructive to mouth and digestive enzymes. Try herbs, parsley, chives; use garlic, leeks, marjorum and special vegetable seasonings—that is, celery salt, garlic salt, onion salt, etc.

Where to buy: Organically grown foods are available at many health food stores as are health seasonings. You may find a farm in your vicinity and buy direct; ask you state agricultural division. Write to *Organic Gardening,* Emmaus, Pennsylvania, for a list of nationwide organic farmers who sell everything from steaks to asparagus; many sell through the mails.

10 Rules for Healthier Cooking

Here is a list of 10 rules that will improve cooking and create masterpieces of culinary arts that are also storehouses of nutrients and enzymes:

1. *White foods have poorer nutrient quality.* Vegetables which have been subjected to chemical bleaches are lower in nutrient store. When purchasing, ask the man to leave on outer leaves. Select yellow rather than white turnips. Choose darker sweet potatoes rather than light. Use brown sugar instead of white; use dark whole wheat grains rather than bleached white flour; always use brown whole rye rather than gray.

2. *Stale foods have few nourishing qualities.* A stale food has less nutrient value. Perishable foods should be speedily consumed. Avoid leftovers. They are *not* fresh foods no matter how much of a master chef you are in imparting a delicious flavor to done-over foods.

3. *Eat it raw.* Each and every day, with just about each and every meal, serve a fresh raw fruit and vegetable salad. Raw foods represent your prime source of enzymes.

4. *Best cooking methods.* Listed in order, the best cooking methods include waterless cooking, steaming without pressure, broiling, pressure-cooking, baking, deep-fat frying, simmering (slow boiling), and shallow-fat frying. Since frying of any type causes an uneven breakage of fat globules within the foods which will coat the digestive apparatus and render enzymes useless, you should avoid this type of cooking. While cooking diminishes nutrient value as well as enzyme power, the above methods are least destructive.

5. *Cook less . . . not too much.* Cook as little as is necessary; overdone and overcooked foods have less nutrient value.

6. *Cook and serve promptly.* Avoid cooking and storing foods for the days ahead. Heat, air exposure and time are destructive to nutrients and enzymes as well as flavor. Cook and serve as soon as possible.

7. *Keep air away from foods.* Air is good for your lungs but will destroy valuable nutrients in foods. When cooking, close pots and pans. Avoid beating air into foods being prepared or cooked.

8. *Soda is taboo.* Baking soda may give vegetables a green color, but it will also give you heartburn and indigestion—baking soda is destructive to enzymes and that accounts for faulty digestion.

9. *Cooking water is healthy.* Save all cooking juices and water. Use it for stews, soups, sauces or as gravies. They are rich in nutrients that have been taken from the vegetables that they were cooked in.

10. *Use vegetable oils.* Wherever possible, use unsaturated fats such as corn oil, cottonseed oil, peanut oil, olive oil, sunflower seed oil, safflower seed oil, wheat germ oil, etc. Most groceries and supermarkets have a wide variety of cooking oils from which you may make a selection.

Your "Instant Health" Handy Cookbook

SUNFLOWER SEED LOAF

1½ cups ground sunflower seeds
¾ cup finely ground sesame seed meal
½ cup chopped walnuts
1 cup cooked lentils
½ cup grated raw beets
3 tablespoons minced chives
2 eggs, beaten slightly
1 tablespoon apple cider vinegar
½ cup diced celery
½ cup whole grain flour
2 teaspoons lemon juice

Blend together all items, press into an oiled baking dish. Bake until done, about 40 minutes, at 325° F. Serve hot from the oven with a raw salad.

RICE WITH CHICKEN—SPANISH STYLE

1 frying chicken cut in
serving pieces
½ cup peanut oil
1 cup uncooked rice
2 cups chicken broth

¼ cup tomato puree
1 small onion, chopped
1 clove garlic, minced
6 green olives
pinch powdered saffron

Preheat oven to 350° F. Saute chicken in the oil to light brown color; transfer to heatproof dish; continue cooking in oven while rice is being prepared. Heat oil in a pan, add rice and stir until yellow. Add onion, garlic, olives, saffron, saute a few minutes. Add hot chicken broth and tomato sauce. Cook gently until all stock is absorbed and rice is almost done. Add more broth if necessary to keep rice moist. Stir occasionally. This will take about 20 minutes. Place chicken pieces on top of rice, cover and continue cooking in oven for 10 minutes. Serve with pumpernickle bread.

VEAL SCALOPPINI

1 lb. veal (cutlets or steak)
¼ cup flour
½ cup apple cider vinegar

4 tablespoons vegetable oil
pinch of vegetized salt
1 sliced lemon

Have veal cut thin and flattened and cut into 4-inch pieces. Roll veal in flour. Heat oil in skillet; brown veal quickly. Add apple cider vinegar. Cover; simmer over low flame about 5 minutes or until meat is tender. Sprinkle with salt and pepper. Serve very hot with lemon slices.

LIVER AND ONIONS—PARIS STYLE

4 large onions
¼ teaspoon vegetable salt
1 teaspoon vegetable oil

1 lb. calves' liver
or baby beef liver
2 tablespoons apple cider
vinegar

Cut onions into small pieces; add salt and saute in oil until golden. Cut liver into 1-inch cubes, discarding skin and veins. Add liver to onions, cooking and stirring a few minutes longer. Apple cider vinegar is added just before serving.

BAKED FISH

(use striped bass, red snapper, carp, mackerel or other fish.)

2 lbs. fish
2 medium sized onions
½ cup tomato puree
1 clove garlic, crushed
6 tablespoons vegetable oil

½ teaspoon vegetable salt
juice of ½ lemon
chopped parsley
lemon slices

Cut onions into slices and saute in 4 tablespoons of oil until golden brown. Add tomatoes, garlic, vegetized salt, lemon juice and remaining oil. Cook for about 20 minutes. Place fish in baking dish and cover. Bake in pre-heated oven at 400° F., approximately half hour. Before serving garnish with chopped parsley and lemon slices.

SKILLET STEAMED FISH

(use fillets of fish, small fish steak or small whole fish.)

1½ lbs. fish
4 tablespoons vegetable oil
1 tablespoon lemon juice

seasonings
whole wheat flour (unbleached)

Season fish with herbs or vegetized salt. Sprinkle with lemon juice. Dip the fish lightly in flour. Heat oil in skillet. Cook fish until golden brown on both sides, about 10 minutes. Serve with tomato slices.

SAVORY SWISS STEAK

1 lb. beef round
herbal seasonings
2 cups tomato juice

whole grain flour
sunflower seed oil

Season meat, sprinkle with whole grain flour. Pounding helps to tenderize. Cut meat into serving pieces and brown in heated oil in a skillet. Add tomato juice, cover and simmer gently until meat is tender, about 90 minutes.

MEAT AND VEGETABLE PIE

1 cup cubed carrots
1 small onion, sliced
1 cup whole wheat
    biscuit dough

1 cup cubed potatoes
1 cup cooked ground meat

Steam vegetables until tender. Combine with meat. Heat thoroughly in pot, then pour into a baking pan. Cut dough into biscuits and arrange on top of meat mixture. Bake at 425° F., about 15 minutes or until biscuits are brown.

BAKED LAMB STEW

1½ lbs. cubed lamb shoulder
2 cups sliced beets
2 cups diced tomatoes
3 cups vegetable bouillon
¼ cup rye flour (unbleached)

1 cup sliced onions
1½ cups cut green beans
1 cup sliced mushrooms
1¼ cups biscuit mix

Combine lamb and onions. Cook over low heat until lamb is browned on all sides. Add beets, green beans, tomatoes, mushrooms, bouillon and seasonings. Mix well. Turn into 3-quart casserole. Cover and bake at 350° F., one hour or until lamb and beets are tender. Combine biscuit mix and rye flour. Add water and mix lightly. Turn out on lightly floured surface and knead gently 10 times. Roll out to ½ inch thickness. Cut into 2½ inch rounds, using flour cutter. Arrange biscuits over stew. Bake at 400° F., about 15 minutes or until biscuits are done.

LAMB FRICASSEE

2 tablespoons salad oil
1½ cups water
3 medium onions, sliced
1½ tablespoons whole
    wheat flour

cooked brown rice
2 lbs. lamb shoulder, cubed
2 cups sliced carrots
mint flakes
1½ tablespoons cold water

Heat oil. Add lamb and cook until lightly browned on all sides. Drain off drippings. Add 1½ cups water, carrots, onions, mint flakes and seasonings. Cover and cook over low heat for 2 hours. Combine flour and 1½ tablespoons water; blend. Add to lamb mixture. Cook over low heat, stirring constantly, until thickened. Serve lamb mixture over rice or with peas.

NEW ENGLAND FISH CHOWDER

1 lb. fish (cod, halibut or
  whiting)
2 tablespoons vegetable oil
2 cups boiling water
1 cup diced raw potatoes

1½ teaspoon vegetized salt
¼ cup chopped onion
2 cups skim milk
chopped parsley

Bring seasoned water to a boil. Add fish. Simmer for 15-20 minutes (do not boil). Cook onions in oil until transparent. Add potatoes, fish stock, boil 15 minutes or until potatoes are tender. Add fish, flaked and with all bones and skin removed. Heat milk and add. Let stand for a few minutes to blend flavors. Sprinkle and serve with chopped parsley.

SWEET AND PUNGENT CHICKEN LIVERS

3 green peppers
¾ lb. chicken livers
1 tablespoon vegetable oil
½ teaspoon celery salt
1 cup chicken bouillon

4 slices pineapple
1 tablespoon whole wheat flour
1 tablespoon soy sauce
¼ cup apple cider vinegar
¼ cup brown sugar

Cut each green pepper into 6 pieces, parboil until nearly tender in boiling, seasoned water, about 5 minutes. Drain. Brown quartered chicken livers in the oil in a heavy 10-inch skillet. Add ½ teaspoon celery salt. When brown, remove to hot serving dish and keep warm. Now put 1/3 cup chicken bouillon, the pineapple and green pepper in the skillet. Cover and cook over very low flame about 10 minutes. Blend together the flour, soy sauce, apple cider vinegar, brown sugar, and the remaining chicken bouillon. Add this to the skillet, cook, stirring constantly until thick, about 5 minutes. Pour over chicken livers and serve at once.

BEEF CASSEROLE

1½ lbs. chuck beef
2 tablespoons millet meal
1 teaspoon vegetized salt
3 tablespoons cooking oil

1 cup tomato puree
1 garlic clove
hot water

Cut beef into one inch cubes; now roll in the millet meal into which you have blended salt. Heat cooking oil in heavy skillet. Brown beef cubes lightly. Pour over them the tomato puree in which garlic clove has been soaking. Add enough boiled water to cover. Bake at 350° F., in a covered casserole for about 90 minutes or until meat cubes are tender.

BROILED LIVER SURPRISE

1½ lbs. calves' liver	1 garlic clove
pinch of mace	4 tablespoons vegetable oil
½ teaspoon onion salt	1 tablespoon lemon juice
1 onion cut in rings	

Cut liver into ½ inch thick slices. Rub both sides with cut halves of garlic clove. Sprinkle lightly with onion salt. Dash minutest bit of mace on one side only. Brush both sides with vegetable oil. Place on broiler rack about 3 inches from the medium flame. Place onion rings on each slice just before putting meat on broiler rack. Broil for 6 minutes, then turn (remove onion rings and replace on other side). Broil for another 6 minutes or until done. Moments before removing, sprinkle lemon juice on each piece of broiled liver. This cautious broiling will seal in all valuable juices to give you a maximum of valuable nutrients.

SHRIMP WITH GREEN PEPPER AND TOMATO

1½ lbs. raw shrimp	½ teaspoon ginger powder
1 lb. green pepper	1¼ cups vegetable stock
1 clove crushed garlic	2½ tablespoons soya sauce
2 tablespoons whole wheat flour	4 tablespoons vegetable oil
1 lb. fresh tomatoes	4 scallions, cut in 1″ long pieces

Mix flour and soya sauce in ¼ cup water and set aside. Cut each pepper diagonally into 8 pieces and parboil for 5 minutes. (Use the cooking water for part of the stock.) Cut tomatoes into wedges, 6 to 8 pieces to each. Shell and split shrimp along the back, but do not cut all the way through. Place oil, seasonings and garlic in hot skillet, add shrimp, saute 3 minutes, add green peppers and stock, mix well and cook gently for about 8 minutes. Add flour, tomatoes and scallions. Continue stirring carefully so that the tomatoes won't break up, until juice thickens smoothly. Serve steaming hot.

MEATLESS ROAST

1 can Protose or Nuttose (8 to 16 oz. can)	4 celery stalks (cut fine)
2 medium onions, diced	few sprigs parsley, chopped
1 large green pepper, diced	1 tablespoon celery salt
1 clove garlic	4 slightly beaten eggs

Grate nutmeat on ordinary coarse grater or grind in food chopper. Mix with other ingredients. Bake in a buttered loaf pan for 20 minutes. Serve with raw salad dish.

CHESTNUT PIE (MEATLESS MAIN DISH)

¾ cup cooked (coarsely
   chopped) chestnuts
2 cups coarsely cut mushrooms
1 tablespoon cooking oil

½ cup cream
1½ cups milk
1 rounding flour

Cook mushrooms for 15 minutes with just enough water to cook dry. Add oil about 5 minutes before they are done. Now add cream and milk and the flour stirred smooth in a little of the milk. Cook up. Add the chestnuts and stir through. Put in a casserole, cover with pastry and bake to a nice brown.

SPINACH CHOPS

1/3 cup spinach puree
1½ cups soya milk

3 eggs
vegetaized salt

Cook spinach tender but not overdone. Chop fine or press through colander. Add to other ingredients, whip up well, pour into an oiled pan, bake in a moderate oven until firm; turn out on a platter, cut into shape or chops, serve with celery sauce.

SOY BEAN CHEESE LOAF

1 14-ounce can soy cheese
1 cup cracker crumbs
½ cup hot soya milk
¼ cup strained tomatoes
2 eggs

½ cup ground English walnuts
1 small onion (grated)
pinch of mace
pinch of celery salt or sage

Rub the cheese, nuts and onion together quite fine. Pour hot soya milk over the crumbs. Now combine all ingredients, except egg whites, together. Whip the whites stiff and fold through mixture lightly. Turn into oiled and crumbed baking dish and steam or bake for 30 minutes.

LIMA BEAN STEAK

3 cups cooked lima beans
1½ cups cubed bread
1¼ cups soya milk
2 teaspoons chopped parsley
¼ cup finely chopped onion

1 tablespoon corn oil
1 teaspoon sage
¼ teaspoon celery salt
2 eggs
1 tablespoon margarine

Braise onion in corn oil until tender but not brown. Whip the eggs; add soya milk and pour over the bread. Then add the beans and all seasonings. Stir through well. You may bake in a casserole, all ingredients combined. Tastes just like steak.

LEEK AND POTATO SOUP

4 small leeks
4 tablespoons vegetable oil
3 medium potatoes

2 cups broth or consomme
2 cups skim milk
seasonings

Finely mince the leeks. Saute them in the oil until transparent for about 15 minutes. Peel and slice potatoes and add. Heat consomme and add. Add herbal seasonings and cook gently until potatoes and leek are tender, about 30 minutes. When about to serve, add hot skim milk. (Note: Shredded green cabbage may be used in place of leek.)

FRENCH ONION SOUP

3 tablespoons vegetable oil
2 large onions, sliced thin
4 cups stock or broth

2 tablespoons grated cheese
seasonings

Saute onions in oil until golden. Add stock. Simmer 30 minutes. Season according to taste. Sprinkle with cheese before serving.

CELERY CHOWDER

1 cup sliced potatoes
1 tablespoon chopped onion
1 cup chopped celery

1 tablespoon butter
½ cup soya milk
2½ cups water

Cook the potatoes, onions, and celery in the water until tender. Add butter and cook up. Now add soya milk. Bring to a boil and serve.

SPLIT PEA VITAMIN SOUP

3 cups split peas
2 bay leaves
4 tablespoons corn oil
1 medium onion (chopped fine)
vegetized salt to taste

3 cups diced potatoes
1 cup chopped celery
(with leaves)
2 cups chopped parsley

Soak peas. Add 3 quarts water and cook together with bay leaves, oil, garlic and onion as well as required vegetized or herbal seasonings. Cook until dissolved, then add potatoes, celery and parsley. Cook slowly until vegetables are tender, about 20 minutes.

EGGS A LA KING

½ lb. mushrooms   1 teaspoon butter or margarine
2 hard-boiled eggs (cubed)   ¼ cup soya milk

Cut medium-size mushrooms in six or eight pieces. Wash in strainer (it is not necessary to peel them), cook in small amount of water 10 minutes, add butter and cook three minutes longer. Now add milk. Remove from fire and add eggs. Serve on whole wheat bread.

STEAMED EGGS

Heat a skillet, add a small amount of cooking oil, break in the eggs, add two tablespoons of water. Cover and allow to cook on low heat until desired degree of hardness has been reached. Use eggs cups, custard cups or small pottery dishes set in water for this type of egg cookery.

NUTMEAT OMELETTE

1 egg   ½ slice any meatless protein loaf
salad oil

Cut meatless protein loaf (Madison, Loma Linda, Worthington, Nuvita, etc.) in round or half-moon slices. Heat on both sides over a low flame in salad oil. Beat egg with a fork, drop on nutmeat and bake under electric broiler or in oven on top of the stove on low flame until egg is set.

VEGEX GRAVY

2 tablespoons chopped onion   1 level tablespoon Savita or
1 tablespoon butter or   Vegex
  margarine   ½ teaspoon celery salt
2 tablespoons corn oil   ¼ cup whole wheat flour

Add chopped onion to butter. Simmer until tender. Add flour, then the Savita or Vegex. Stir through well. Now add enough water to make desired consistency and seasonings. Cook up. Use this gravy for any meat or meatless dish.

PARSLEY SAUCE

1½ tablespoons vegetable oil   1 cup hot soya milk
2 tablespoons rye flour   2 tablespoons chopped parsley
¼ cup cold soya milk

Add rye flour to vegetable oil in saucepan; dissolve over slow fire. Add cold milk and stir through. When it begins to thicken, add the hot milk. Whip smooth and boil. Add parsley. A delicious sauce for any accompaniment.

ALMOND CREAM GRAVY

3 tablespoons margarine
1/3 cup sliced almonds
2 tablespoons flour

2 cups soya milk
celery salt

Saute almonds in melted margaine until light brown. Stir in flour; when blended, add milk and celery salt. Stir until smooth and the desired consistency. Put in gravy boat and serve piping hot.

EGGPLANT ROAST

1 medium size eggplant
(prepare as for broiling)
3 sliced onions

2 sliced green peppers
1 cup tomato juice

On bottom of pan, put ½ of the sliced onions and ½ sliced green peppers. Place a layer of round eggplant slices. For next layer, use remaining half of onions and green peppers. Top with remaining eggplant slices. Put a slice of tomato on each piece of eggplant or pour tomato juice over mixture. Cover pan with tight lid. Bring to a boil and simmer for 20 minutes. Season before serving with salad oil.

WHOLESOME MAYONNAISE

1 whole egg and 1 yolk
1 pint sunflower seed oil

2 tablespoons lemon juice

Break eggs in bowl. Add lemon juice and beat up. Add sunflower seed oil slowly at first, then faster, until all oil is used. Should take about 5 minutes. Use an egg whip. Serve on lettuce or raw vegetable salads.

CELERY DRESSING

½ cup wholesome mayonnaise
(see above)

1 tablespoon chopped green pepper
¼ cup chopped celery

Mix all together. Serve on plain lettuce.

SOUR CREAM DRESSING

½ cup of sour cream
¼ teaspoon celery salt

2 tablespoons powdered brown sugar
2 tablespoons lemon juice

Whip the cream, season with salt, sugar and lemon juice. You may use sweet cream for a substitute; add extra lemon juice for more tang.

LEMON MAYONNAISE

1 egg yolk   1 cup cooking oil
vegetized salt   3 tablespoons lemon juice

Add salt to the egg yolk. Beat until it is creamy. Add the oil drop by
drop, beating it well after each addition. Continue this until all the oil
is added and the mixture is thick. Beat in the lemon juice.

HOME MADE FRUIT DESSERT

1 cup whipping cream   1/3 cup chopped nutmeats
3 tablespoons honey   1 cup mixed seasonal berries

Whip the cream and blend in the honey. Mix well with the fruit, then
stir in nutmeats. Chill. Serve in sherbet glasses, topped with a fresh
strawberry.

STARCH-FREE COOKIES

1 cup crushed walnuts or pecans   ½ cup honey
4 egg whites   1 cup dry cereal
⅛ lb. butter

Beat egg whites until stiff. Mix all dry ingredients individually with a
spoon. Fold in egg whites. Drop by teaspoons onto hot oiled cookie
sheet and bake at 350° F. for 20 minutes. (Instead of nuts, use dates,
raisins or chopped figs.)

Your Two Week "Instant Health" Diet Plan

FIRST WEEK

MONDAY

BEFORE BREAKFAST: Special enzyme cocktail (Chapter 4.)
BREAKFAST: Half cantaloupe with cottage cheese; soft boiled egg with
Melba toast; cup fresh yogurt with prune puree; coffee substitute.
MID-MORNING STIMULATOR: Jet enzyme tonic (kefir drink, Chapter
15.)
LUNCHEON: Salad of sliced tomato, lettuce, celery, cucumbers with
dressing of equal parts of safflower seed oil and apple cider vinegar,
broiled liver patty; side dish of soya noodles; fenugreek tea with lemon
juice and honey.
MID-AFTERNOON PICKUP: 1 glass freshly squeezed pineapple juice.
DINNER: Steamed brown rice, broccoli; escarole, carrot slices, radishes;

broiled whitefish with protein bread; dish of figs; rose hips tea with lemon juice and honey.

TUESDAY

BEFORE BREAKFAST: Special enzyme cocktail.
BREAKFAST: Bowl of oatmeal with spoon of blackstrap molasses; banana and sour cream; six de-pitted raw prunes; coffee substitute, or glass of soya milk.
MID-MORNING STIMULATOR: Jet enzyme tonic.
LUNCHEON: Shrimp salad on bed of lettuce leaves; raw green peas and celery; fruit flavored gelatin; alfalfa tea with lemon juice and honey.
MID-AFTERNOON PICKUP: 1 glass freshly squeezed peach juice.
DINNER: Rice with chicken—Spanish style (see recipe); pumpernickle bread; slices raw onion, crisp celery; broccoli and steamed cabbage; several whole wheat cookies; maté tea with lemon juice and honey.

WEDNESDAY

BEFORE BREAKFAST: Special enzyme cocktail.
BREAKFAST: Raw fruit plate of sliced plums, berries, grapes; French toast; dish of raisins; coffee substitute.
MID-MORNING STIMULATOR: Jet enzyme tonic.
LUNCHEON: Sardine sandwich with whole grain bread; beets, radishes, raw green pepper slices; carob candy bar; sarsaparilla tea with lemon juice and honey.
MID-AFTERNOON PICKUP: One glass freshly squeezed apple juice.
DINNER: Sunflower seed loaf (see recipe); salad of shredded raw cabbage, cauliflower, parsnips; protein bread; glass of soya milk into which is stirred 3 tablespoons of tupelo honey.

THURSDAY

BEFORE BREAKFAST: Special enzyme cocktail.
BREAKFAST: Poached egg on rye bread; peach halves with raisins in bowl of soya milk; coffee substitute.
MID-MORNING STIMULATOR: Jet enzyme tonic.
LUNCHEON: Baked eggplant; bowl of chopped, hard-cooked egg, cubed tomatoes, chopped spinach, chopped olives; whole wheat rolls; glass freshly squeezed carrot juice with Brewer's yeast flakes.
MID-AFTERNOON PICKUP: 1 glass freshly squeezed cabbage juice.
DINNER: Savory Swiss Steak (see recipe); diced celery, avocado, green pepper; protein bread; honey cake; parsley tea with lemon juice and buckwheat honey.

FRIDAY

BEFORE BREAKFAST: Special enzyme cocktail.
BREAKFAST: Steel cut corn flakes in soya milk; sliced apples, pears, de-pitted dates; cup of yogurt with Brewer's yeast flakes and tupelo honey; coffee substitute.
MID-MORNING STIMULATOR: Jet enzyme tonic.
LUNCHEON: Natural Amercian cheese sandwich on whole grain bread with succulent lettuce and tomato; dish of grapes, figs, sliced bananas; fruit flavored gelatin; sarsaparilla tea with lemon juice and clover honey.
MID-AFTERNOON PICKUP: 1 glass freshly squeezed beet and lettuce juice.
DINNER: Raw salad of endive, onion slices, sprinkled with ground walnuts; skillet steamed fish (see recipe); pound cake; red clover tea with lemon juice and honey.

SATURDAY

BEFORE BREAKFAST: Special enzyme cocktail.
BREAKFAST: Wheat flakes with raisins in soya milk; slices of apples, pears, plums; oatmeal cookies; 1 cup hot water with 4 tablespoons black-strap molasses.
MID-MORNING STIMULATOR: Jet enzyme tonic.
LUNCHEON: Medium baked potato with jackets; cup green beans and asparagus tips; baked codfish with lettuce and tomato; oatmeal cookies; sarsaparilla tea with lemon juice and honey.
MID-AFTERNOON PICKUP: 1 glass freshly squeezed celery juice.
DINNER: Baked lamb stew (see recipe); salad of raw spinach, shredded cabbage; green peppers and asparagus; dietetic pudding; huckleberry tea with lemon juice and honey.

SUNDAY

BEFORE BREAKFAST: Special enzyme cocktail.
BREAKFAST: Soft boiled egg with whole wheat bread slices; 3 bananas with side dish cottage cheese; coffee substitute.
MID-MORNING STIMULATOR: Jet enzyme tonic.
LUNCHEON: Natural Swiss cheese on protein bread with chopped pimiento and parsley; cup of soya milk with powdered wheat germ; maple sugar candies.
MID-AFTERNOON PICKUP: 1 glass of coconut juice.
DINNER: Meat and vegetable pie (see recipe); dish of steamed green peas and squash; raw salad of lettuce, radishes, green peppers; whole wheat bread with soy lecithin spread; fennel tea with lemon juice and honey.

SECOND WEEK

MONDAY

BEFORE BREAKFAST: Special enzyme cocktail.
BREAKFAST: One egg yolk stirred into freshly squeezed orange juice; puffed wheat cereal in yogurt; banana; coffee substitute.
MID-MORNING STIMULATOR: Jet enzyme tonic.
LUNCHEON: Cashew nut butter sandwich on whole wheat bread; celery, carrot and tomato slices; side dish of soya spaghetti; peppermint tea with lemon juice and orange blossom honey.
MID-AFTERNOON PICKUP: 1 glass cucumber-celery juice.
DINNER: Lamb fricassee (see recipe); broccoli, fresh corn, cauliflower; rye bread; fruit cake slice; camomile tea with lemon juice and honey.

TUESDAY

BEFORE BREAKFAST: Special enzyme cocktail.
BREAKFAST: Whole wheat waffles with maple syrup; seasonal berries with sliced raw figs; coffee substitute.
MID-MORNING STIMULATOR: Jet enzyme tonic.
LUNCHEON: Raw vegetable sandwich of chopped olives, radishes, cucumbers, cabbage; small cup mushroom soup; dish of steamed brown rice; papaya-mint tea with lemon juice and honey.
MID-AFTERNOON PICKUP: 1 glass freshly squeezed cherry juice.
DINNER: Sweet and pungent chicken livers (see recipe); beet greens, yellow squash, okra, raw celery and carrots; whole wheat or rye bread; low-calorie ice cream; juniper berry tea with lemon juice and honey.

WEDNESDAY

BEFORE BREAKFAST: Special enzyme cocktail.
BREAKFAST: Soft boiled egg on soya bread; seasonal berries in sour cream; raw raisins and sunflower seeds; coffee substitute.
MID-MORNING STIMULATOR: Jet enzyme tonic.
LUNCHEON: Natural muenster cheese sandwich on protein bread; salad of lettuce, shredded cabbage, swiss chard; gelatin dish; fenugreek tea with lemon juice and honey.
MID-AFTERNOON PICKUP: Have a mixed vegetable drink.
DINNER: Broiled liver surprise (see recipe); raw turnip slices, asparagus, mixed nuts; honey cake; goldenrod tea with lemon juice and honey.

THURSDAY

BEFORE BREAKFAST: Special enzyme cocktail.
BREAKFAST: Dish of apricot slices in soya milk with sprinkles of sesame seeds; portion of de-pitted prunes and grapes; fresh banana; coffee substitute.
MID-MORNING STIMULATOR: Jet enzyme tonic.
LUNCHEON: Broiled hamburger on whole wheat bun; onion rings with raw shredded cabbage and spinach; blueberry muffins; papaya mint tea with lemon juice and honey.
MID-AFTERNOON PICKUP: Have a glass of pomegranate juice.
DINNER: Beef casserole (see recipe); diced carrots, beets, endive; dietetic pudding; rose hips tea with lemon juice and honey.

FRIDAY

BEFORE BREAKFAST: Special enzyme cocktail.
BREAKFAST: Whole wheat pancakes with butter pat; dish of raw figs sprinkled with wheat germ; coffee substitute.
MID-MORNING STIMULATOR: Jet enzyme tonic.
LUNCHEON: Steamed scallops with hot buttered rolls; diced beets, turnips, cabbage; cinnamon buns; camomile tea with lemon juice and honey.
MID-AFTERNOON PICKUP: Have a glass of celery juice.
DINNER: Eggplant roast (see recipe); lima beans and corn; raw celery, escarole, Chinese cabbage; dietetic gelatin with yogurt scoop; shave grass tea with lemon juice and honey.

SATURDAY

BEFORE BREAKFAST: Special enzyme cocktail.
BREAKFAST: Soft boiled egg; protein bread with dietetic jam; dish of seasonal berries and raisins; coffee substitute.
MID-MORNING STIMULATOR: Jet enzyme tonic.
LUNCHEON: Sardine sandwich on rye bread with lettuce and tomato slices; small cup thick tomato and rice soup; huckleberry muffins; maté tea with lemon juice and honey.
MID-AFTERNOON PICKUP: 1 glass sauerkraut juice.
DINNER: Lima bean steak (see recipe); celery slices, raw green peas, cauliflower; dish of dates and mixed nuts; sarsaparilla tea with lemon juice and honey.

SUNDAY

BEFORE BREAKFAST: Special enzyme cocktail.
BREAKFAST: French toast with soya butter; yogurt with prune puree and wheat germ flakes; fresh apple; coffee substitute.
MID-MORNING STIMULATOR: Jet enzyme tonic.
LUNCHEON: Beef hamburger on whole wheat bun with sliced onion ring and cucumber; dish of sliced tomatoes, celery, carrot strips; starch-free cookies (see recipe).
MID-AFTERNOON PICKUP: 1 glass cabbage-spinach juice.
DINNER: Meatless roast (see recipe); dish of chard, broccoli, kale; soya bread; dietetic butterscotch pudding; fenugreek tea with lemon juice and tupelo honey.

Enzyme Location Map at a Glance

It is estimated that more than 20,000 chains of enzymes exist in the body so it would be impossible to list all of them; science has not yet isolated even a fraction of enzymes, let alone identify them. Here are some of the more known and pertinent *endogenous enzymes* (those found within your body):

ENZYME	WHERE FOUND	FUNCTION
Ptyalin	Mouth Saliva	Turns starch into energy-producing sugar
Pepsin	Stomach	Acts upon proteins
Rennin	Stomach	Acts upon milk foods
Lipase	Stomach	Acts upon fats in foods
Trypsin	Intestine	Works on protein foods
Streapsin	Intestine	Assimilates fat foods
Amylopsin	Intestine	Assimilates starch foods
Invertase	Intestine	Attacks sugar in foods
Maltase	Intestine	Assimilates malt sugars in foods
Lactase	Intestine	Works on milk sugars in foods
Rennin	Intestine	Further acts upon milk foods
Erepsin	Intestine	Converts proteins into usable amino acids
Phosphatase	Bloodstream	Builds phosphorus into system
Sucrase	Hormonal	Changes sugar into usable sucrose
Hydrochloric Acid	Gastric	Works upon tough fibrous foods to take out amino acids
Thyroxin	Thyroid gland	Mental stability and hormonal balance
Chymotrypsin	Pancreas	Reduces inflammatory swellings

Exogenous enzymes (Found in raw foods):

ENZYME	FUNCTION
Urease	Breaks down urea, substance formed in the liver and delivered to kidneys via bloodstream.
Papain	Aids in digestion of protein foods; found in the papaya.
Cellase	Attacks indigestible cellulose of tough and fibrous foods such as celery, carrots, root vegetables.
Fibrinolysin	Dissolves strong fibrinous deposits and blood clots.
Penicillinase	Eases allergic symptoms to the penicillin antibiotic.
Chymotrypsin	Dissolves anchor fibers in cases of visual cataracts.
Procollagenase	Will "erase" disfiguring scars and intestinal adhesions after surgery; dissolves material from which tough fibers develop.
Elastase	Secreted by pancreas but also isolated and seen to dissolve cholesterol deposits in arteries.
Streptokinase	Dissolves blood clots and reduces swellings.
Streptodornase	Absorbed into bloodstream and also dissolves clots.
Bromelin	Found in pineapple stems; seen to ease muscular tensions and female problems.
Papain	In addition to above mentioned function, this enzyme dissolves obstructed tissues of a slipped disk.
Histidase	Influences memory and mental stability-intelligence.
Dopamine	Stabilizes nerve ganglia and brain inflammation condition of Parkinson's disease (shaking palsy).
Sorotonia	Same as above, dopamine.
Cholinesterase	Relieves condition of schizophrenia.
Catalase	Builds resistance against cancer.

(NOTE: Raw fruits and vegetables, raw seeds and nuts, etc., will provide you with a tremendous supply of these exogenous enzymes. However, scientists, researchers and doctors are constantly isolating them, as per a few listed in the chart, and they are applied and administered by means of injection. Some enzymes are available in a highly concentrated form by a physician's prescription. Consult with your family doctor who is the best source for advice on obtaining the newly discovered exogenous enzymes.)

A Few Words in Conclusion

"The work that I was born to do is done," wrote a great poet when he reached the completion of his task. And while I do not feel entitled to sing any *Nunc dimittis,* I am fully aware that the task has taken up the major portion of my life.

I am also cognizant of the attitudes of many who will read the discoveries in this volume. Some will consider these truths to be too conservative; others will consider them to be too overwhelming. It is to be expected that we always have some who passionately seek to cling to the past; there are just as many who passionately seek to snatch at what they believe to be the future. It is the wise person, standing between both parties, sympathizing with each, who understands that we are always in a transitional stage.

The present is little else than the intersection at which past and future meet. We cannot dispute either; there could be no life without past traditions, neither could there be life without future movements.

The wise Grecian health-philosopher, Heraclitus, at the dawn of civilization, once opined that we cannot bathe twice in the same stream, although the stream flows in an unending circle. There is never a single second when a brand new Dawn does not break over the earth; there is never a single second when the Sunset does not descend upon us. A wise course is to welcome the earlier glimmers of Dawn when we view it; we need not hasten toward it for we know that Sunset will eventually come.

Today, we become the bearers of light since Dawn and Sunset have given us the beginning and the end; or, the past and the future. We, therefore, must stand in the middle of the ever flowing stream of life and examine both directions. In so doing, we may bring light to whatever darkness may obscure our quest for eternal health and eternal youth.

To Lucretius, another Grecian philosopher, the ancient torch-race was symbolic of youthful health. We press forward, the torch of health in our hand. We run along the course. Soon, from behind, another will come to outpace us. We must, therefore, present him with this living torch of health, bright and unflickering, so he may bring light to the many others in the stream of life who seek eternal youth and health.

INDEX

Books on Health, Nutrition

Vitamin E—Your Key to a Healthy Heart
Herbert Bailey

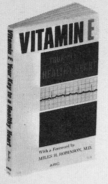

WHY IS VITAMIN E therapy for mankind's foremost killing disease still controversial in the United States? This is one of the questions asked and answered in this slashing, fully documented book. It tells how the efficacy of Vitamin E in the treatment of cardio-vascular disease was discovered by Dr. Evan Shute of Canada, and of the remarkable cures effected by him and his brother, also a doctor . . . how the author himself suffered a severe heart attack and how in a short time he was restored to normal, active life by massive doses of the vitamin . . . how a barrier against Vitamin E has been erected in this country by the medical traditionalists of the American Medical Association at the same time that it is being widely used with spectacular results in such medically advanced countries as England, Germany, France, Italy, and Russia . . . how continuing study indicates that Vitamin E may be an effective preventative for a variety of other diseases.

HERBERT BAILEY is a veteran medical reporter who has informed the American public about other epochal medical discoveries before they were accepted and used.

224 pages   **#1514 paper: $1.65**

THE SOYBEAN COOKBOOK   **Dorothea Van Gundy Jones**
Over 350 kitchen-tested recipes for using the versatile soybean in the family menu from salads and souffles to meat replacement main dishes and desserts. "An interesting and valuable book . . . delightful foods can be made quite rich in protein and very palatable as well."—Health Culture.

256 pages   **#1770 paper: $1.45**   **#1805 cloth: $3.50**

Food Facts and Fallacies
The Intelligent Person's Guide to Nutrition and Health
Carlton Fredericks, Ph.D. and Herbert Bailey

THIS IS a fascinating, excitingly readable book full of the essential facts that every modern thinking person should know about what he eats, what he should eat, and health. Dr. Carlton Fredericks, for the past 25 years one of our most eminent experts on nutrition has a devoted following which numbers in the millions. This informed group has had the benefit of his unbiased holistic approach to nutrition and his dedication to preventative medicine. They reject the cliche that "ignorance is bliss" or that it is "smart to be gullible." In what is probably the most comprehensive work in print on this

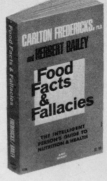

subject for the layman, Dr. Fredericks (with his co-author Herbert Bailey) amplifies and clarifies many previously tested discoveries and much that only recently has found its way into medical and bio-chemical scientific research. Herbert Bailey, his co-author, is a pioneer whose many books and magazine articles have spearheaded recognition of medical discoveries which might otherwise be lost to public use. Never before has there appeared in a single volume for the general public so much real information, so complete a picture of the latest research on nutrition and health as will be found in **FOOD FACTS AND FALLACIES**.

384 pages   **#1726 paper: $1.45**

ARTHRITIS CAN BE CURED   Bernard Aschner, M.D

If you are one of the millions of arthritis victims or if a friend or someone in your family suffers from an arthritic disease, this book can bring new hope into your life. Arthritis is officially considered to be an intractable and almost incurable malady. The most widely accepted treatments, at best, are long-term and slow to bring results; at worst, pointless or even harmful. However, despite the almost universally pessimistic attitude toward the disease, the author of this book, Dr. Bernard Aschner, maintains that arthritis really can be cured; painful, prolonged suffering can be eliminated; incapacitated persons can return to active, useful occupations. And he explains, in clear, easy-to-understand language for the layman, exactly how this may be done.

260 pages   **#1764 paper: $1.45**

Get Well Naturally

Linda Clark

LINDA CLARK believes that relieving human suffering and obtaining optimum health should be mankind's major goal. She insists that it does not matter whether a remedy is orthodox or unorthodox, currently praised or currently scorned in medical circles—as long as it works for you. Mrs. Clark, who is also the author of **Stay Young Longer**, makes a plea for the natural methods of treating disease —methods which do not rely on either drugs or surgery. Drawing not only from well-known sources but also from folklore and from the more revolutionary modern theories, she presents a wealth of in-

formation about diseases ranging from alcoholism to ulcers. Here are frank discussions of such controversial modes of treatment as herbal medicine, auto-therapy, homeopathy, and human electronics, plus some startling facts and theories about nutrition and about the natural ways of treating twenty-two of the most serious illnesses that plague modern man.

410 pages   **#1762 paper: $1.65**

INTERNATIONAL VEGETARIAN COOKERY   Sonya Richmond

This book proves that vegetarian cookery, far from being dull and difficult to prepare, can open up completely new and delightful vistas of haute cuisine. Miss Richmond, who has traveled throughout the world, has arranged the book alphabetically according to countries, starting with Austria and going through to the United States. She gives recipes for each country's most characteristic vegetarian dishes and lists that country's outstanding cheeses.

192 pages   **#1510 cloth: $3.75**
                                                 #1483 paper: $1.75

THE BOOK OF SALADS   Hyla Nelson O'Connor

Here are over two hundred recipes for gourmet salads that are both healthful and delicious. The author also gives valuable information on selecting and handling salad greens and includes thirteen recipes for basic, nutritious salad dressings with many variations.

144 pages   **#374 cloth: $3.50**

HEALTH, FITNESS, and MEDICINE BOOKS

All books are available at your bookseller or directly from ARCO PUBLISHING COMPANY INC., 219 Park Avenue South, New York, N.Y. 10003. Send price of books plus 25¢ for postage and handling. Sorry, no C.O.D.